PERMISSION TO LAND

BRIAN CASHINELLA
and
KEITH THOMPSON

London ⌒ Arlington Books

PERMISSION TO LAND
First Published 1971 by
Arlington Books (Publishers) Ltd
38 Bury Street St James's
London SW1.

Made and Printed in England by
The Garden City Press Limited
London and Letchworth SG6 1JS

© *Brian Cashinella 1971*

ISBN 0 85140 194 5

Permission to Land

CONTENTS

LIST OF ILLUSTRATIONS

[between pages 72 and 73]

ACKNOWLEDGEMENTS

We would like to give our thanks to the following people who gave valuable assistance in the production of this book: A. J. McIlroy, John Clare, Christopher Warman, Frank Robson and the managements of The Times, Daily Express *and* Daily Telegraph, *for allowing us to use photographs from their files. But our special thanks go to John Austin without whose patience and perseverance this book would never have been completed.*

FOREWORD

The object of this book is not to present an academic or wholly technical picture of the complexities involved in choosing the right site for London's third international airport. We feel strongly that the general public, the people who will pay the final bill, have been confused enough.

For years, they have listened to the politicians and so-called technocrats drone on about what they should accept "in the national interest." They have also heard the loud and possibly greedy voice of commercial interest. Often the one has been in direct conflict with the other. The end result, in people's minds, has been utter bewilderment. To twist a cliché, the trees had disappeared into a dense wood.

So, we have unashamedly defoliated the area and attempted to reveal the heart of the matter, principally through the people and personalities involved in the struggle. In this way, we think the reader will discover what really happened, what the issues were and how they were resolved.

It was not an honourable fight. Sometimes, in fact, it became downright dirty, though this was not surprising in many ways. The stakes were high. Political and professional reputations were in jeopardy. Even Governments felt uneasy.

But the long struggle proved one thing: protest, even against a powerful Government machine and its Whitehall servants, is not useless. The "expert" planners were challenged, proved wrong, and forced to relent. Had the preservationists at Stansted (the first choice for the new airport) accepted the Government decision without a battle, the bulldozers would, this very day, be tearing apart a beautiful corner of rural England.

Everyone is now aware that the airport is to be built at Foulness, on the Essex coast. But who swayed the balance? Was it the environmentalists? The City and its commercial partners? Was it the right site in the beginning after all? We leave it for the reader to judge.

Or maybe the Government, of whatever party, would have been forced, in the circumstances, to bend with the prevailing electoral wind—so fickle—and build the new terminal at Foulness. It could have proved politically too dangerous to contemplate otherwise.

We feel it is pertinent here to mention Professor Colin

Buchanan, the eminent town planner, whose sole dissenting voice was heard above all other members of the ill-starred Roskill Commission. Unlike his colleagues who selected Cublington in Buckinghamshire's picturesque Vale of Aylesbury, he insisted that Foulness was the only realistic choice.

Overnight he became a hero. The people of Cublington even sang his praises in their eulogistic hymn for salvation from the persistent roar of overhead jets. He was, they decided, the one sensible man on the Commission and said as much at their public meetings and demonstrations which were to follow. Their attitude was not surprising.

He was the only official defender of their rural way of life, solely by virtue of his convictions about planning priorities. But their gestures were purely emotional, based largely on vested interests. Farms are hard to work, but harder to leave, especially when they have been in the family for generations.

In our investigations, we have always tried to keep in mind the many groups who genuinely wanted and enthusiastically canvassed for, the Government to select an inland airport site. More than 30 unions, for instance, campaigned vigorously on behalf of Thurleigh in Bedfordshire.

Indeed, it has often been suggested that a census of those affected by any airport proposal would result in a resounding victory for the people in favour. But their attempts at presenting their arguments in an articulate and professional public relations manner were miserable failures. The reasons were simple: they were amateurs without high-powered local influence, expertise or money.

Public relations, however, do work as the Cublington campaign proved conclusively. Money talks. Most of the Save Cublington funds came, we were told, from small donations, from ordinary householders, pensioners, organizations like The Friends of the Vale of Aylesbury. Not from big business concerns, at least directly.

One thing, however, is certain. The delays in building the third airport have cost a fortune. Costs of labour, materials, allied to inflation, will probably have doubled the price within a decade. It is to be hoped that there will be no further delay in at least making a start on building the new terminal.

Finally, a personal submission: Britain's tight libel laws have impinged on our freedom to put some people in what we feel to be their proper place. Parliamentary privilege is a wonderful, but often abused law, we think. This harness on journalistic freedom is something, as working newspapermen we have learned to accept. But we feel that even in exposing little more than the tip of the iceberg, the reader may become aware of the inherent dangers that may lie deep under the surface.

BRIAN CASHINELLA

CHAPTER ONE

The Victory

Airports are fascinating for children who like to watch the jets screaming down concrete runways to rise like majestic birds for some faraway romantic paradise.

But they make pretty poor neighbours, especially if you live at the bottom of the runway. Sleep is almost impossible, teaching becomes a frustrating task and clothes hung out to dry are filthy within minutes.

So it was hardly surprising that the rural communities around Stansted picked up their pitchforks in protest when an anonymous Whitehall committee decided the third London international airport should be built at the end of their gardens.

After a fight lasting several years, the farmers won their battle and the selection procedure started all over again, this time via a unique Commission under Mr Justice Roskill.

It is now evident that Roskill was a complete waste of time and money: two years, three million words of evidence and £1,200,000 of the taxpayer's money. For, with the exception of Prof Colin Buchanan, all its members voted in favour of Cublington in Buckinghamshire. That, according to all the rules, should have been that.

However, the people living around this second proposed site did not agree. They were not going to stand idly by and watch their beautiful Vale of Aylesbury ripped apart and turned into a twisting jungle of concrete.

They had watched the success of the Stansted preservationists and decided to launch their own campaign. If Stansted could beat the bowler-hat brigade, then so could Cublington.

But while the brilliant strategy of the Essex farmers was based on saving their land by proving the Whitehall planners had got their sums wrong and done insufficient homework, the Buckinghamshire protesters adopted a different technique.

They appealed largely to the average Briton's ingrained love of the countryside. It was a campaign based on emotions, but carefully controlled, and directed by a clever team of professional public relations men working for a high-powered American-based company called Burson Marsteller.

For a fee of £3,500, the PROs chatted up selected Fleet Street journalists, the television companies, Members of Parliament, to create the "right image" of Cublington: the area became an irreplaceable slice of England's green and pleasant land where an airport, any airport, would be a rape of the countryside.

This professionalism was, of course, backed up with well organised demonstrations by the airport resistance association's 65,000 members. They trundled round the narrow lanes on Sundays with their mud-caked tractors, combine harvesters, cattle trucks and horse boxes—and large stories and pictures appeared in the Press.

It paid off and, once again, the public won the day. The Government decided that the Roskill Commission was wrong, the militant farmers were right and the airport would, after all, go to Foulness—one of the three short-listed alternative sites rejected by the Commission.

In many ways it was an almost unbelievable situation: two successive Governments, one Labour, the other Conservative, bowing to public protest over what has been described as "the biggest single planning decision this country will ever make".

It was a remarkable victory for public opinion and living proof that real democracy still exists, even in a society increasingly controlled by the central administration.

From Foulness, there was only a whimper of opposition, too muted for the Government to consider it would have another fight on its hands. There were the inevitable pockets of resistance; from the Wildlife Trust worried about the future of the birds which used Foulness as a sanctuary; from the 300 residents of Foulness island, most of them pensioners, who were bewildered by the speed of events; from the military who would lose their precious artillery testing ranges at nearby Shoeburyness, a site they have occupied since 1805.

But, unlike either of the other two sites, there was much support for a Foulness air terminal. The local authorities were all in favour and saw the airport as a new industry in an area hit by unemployment; a splendid acquisition for a region with an average weekly wage of under £14.

There would be thousands of well-paid jobs available at the airport itself, they reasoned. Other industries would inevitably follow and this much-neglected area of south-east England would once again taste prosperity. In these circumstances, who cares about noise!

But, in retrospect, the battle for Foulness, for a coastal site airport with its many built-in advantages, was already partially won when a new word entered politics: Environment.

And when Mr Heath, the Prime Minister, appointed the youthful Mr Peter Walker as Secretary of State for the Environment, it provided the new man in the Cabinet with an immediate challenge: here was a fight he had to win.

If he could not persuade his colleagues that an airport, the biggest wrecker of any environment, should be built in a place where it would do no damage, he might as well return to the back benches of the Commons and listen to the noise of the Jumbo jets as they roared overhead towards Heathrow airport.

There was already much support for Foulness in the Cabinet, particularly since private enterprise in the shape of the City's merchant bankers, were prepared to finance the scheme. No Conservative government could afford to ignore that.

There was also the prospect that, given a coastal airport on reclaimed land, a deep-water seaport would be built alongside. And, whilst an airport would be almost non profit-making, in City terms, a seaport certainly would make money—and lots of it. The bankers, and commercially-oriented MPs, had been gazing for the last few years at the remarkable success of Rotterdam's Europoort, just across the North Sea.

They saw no reason at all why Britain should not do the same, especially since our own ports are largely out of date and incapable of taking any of the large container ships, tankers and bulk carriers that are being produced.

So here was the key: build an airport and fast motorway and rail links to central London and you had a ready-made base for a massive seaport complex, and a profitable industrial infrastructure.

Two groups, consortiums of big, international companies, saw the possibilities of such a scheme and formed their own firms to promote the Foulness idea. They provided a powerful force, not only because of their political influence, but because they were prepared to invest huge sums of their own money.

They even offered to build the airport free of charge if the Government would allow them to build their seaport. They considered the building costs of an airport would soon be recouped by the money from rents, fees, warehousing and other charges to ships using the port.

This sort of reasoning weighed heavily with the Cabinet and provided Mr Walker with plenty of ammunition and a strong line of argument. When the Government finally agreed to the Foulness scheme, the young minister was a happy man. He had won his first big battle and his newly-created job took on a new significance.

Also pleased with the situation was Mr John Davies, the Secretary of State for Trade and Industry, who, as a brilliant businessman, saw the commercial viability of developing the coastal area north of Southend.

The formal announcement was made in the Commons on Monday, 26 April 1971, although the Cabinet had paved the way by unofficially "leaking" the decision to the political correspondents at Westminster.

And so, the fight that had started more than a decade earlier, was finished. A site for London's controversial third airport had finally been chosen. The countryside, at one time threatened, was saved.

Not unexpectedly, one man was rather more pleased than most: Mr Bernard L. Clark, the Westminster consulting engineer, who had suggested a scheme for a reclaimed airport site at Foulness in June 1966, was delighted. His unswerving confidence in his proposal had been justified, although some had laughed when the plan was first outlined in *The Daily Telegraph*.

Two problems now remain for the Government. The first is to build an airport with four runways capable of taking the biggest jets, and with facilities that will rival any in the world. The second, and more difficult task, will be to persuade the airlines—and their passengers—to use it. Many people still feel that Foulness, 52 miles from central London, is too far out for a major terminal.

The airlines, already in financial difficulties, are loath to move their base of operations from Heathrow or Gatwick, and still need convincing the extra cost will be worth the effort.

But the protracted battle for the airport will go down as

one of the most fascinating sagas in British aviation history. It will be remembered as the era when the country folk from small farming villages took on the might of Whitehall on two separate occasions—and won both times.

It was a victory for the people, who raised their voices in loud protest and demanded a hearing; who saw the schemes to take over their farms and fields and turn them into runways and hangars as a monstrous piece of social injustice; who saw and heard politicians and airport planners contradicting themselves almost daily in the Press and on television; who were told lies, given false promises—and false hope.

CHAPTER TWO

The Stansted Scandal

It was in the warm midsummer of 1953 that the Ministry of Civil
Aviation, in all its wisdom, decided to rationalise the complications of the London airports situation. At that time the capital
was served by no less than seven different terminals, from Black-
bushe in the west, to Gatwick in the south and Stansted in the
north east. But more than 90 per cent of the 161,000 air movements were at London Airport, and Northolt, which almost sat
on its back door step only five miles away.

From the air traffic control point of view, successfully operating both was becoming not only extremely difficult but downright
dangerous. There was nothing for it but to close Northolt for
civil flying and return it to the Royal Air Force for restricted
military use. Clearly, an alternative would have to be found and
the Ministry, having "considered a large number of aerodromes
and sites within a radius of about 50 miles of central London"
selected Gatwick, because this "alone substantially met all
requirements . . ."

Its main role, according to the White Paper which outlined
the proposal, would be to take some of the seasonal services to
places like the Channel Islands and the near continental resorts,
and to cope with aircraft diverted from London Airport in bad
weather conditions. Furthermore, until the expansion of London
Airport and the development of Gatwick had been completed,
another airport was needed as first reserve "in case traffic increases
even more than is expected". They picked on Stansted but made
it patently clear that it would "cease to be used as a civil airport
once Gatwick is sufficiently developed".

In giving details of the principal sites it had considered, the
White Paper dismissed the case for Stansted in one six-line para-

graph. But, in view of what happened later, those 86 words are particularly relevant: "like Bovingdon, Stansted is on the wrong side of London for most of the aircraft routes. Furthermore, although it is a good aerodrome and would need relatively little expenditure for use as an alternative it has two further substantial disadvantages. Its access by road or rail to London is inferior, travelling time being about one and three-quarters to two hours. There is at present a very large amount of military flying from nearby airfields which severely limits the use which can be made of it by civil aircraft."

While the news, not unnaturally, angered the residents around Gatwick, equally, it cheered the small rural communities around Stansted and nearby Bishop's Stortford, who looked forward, as they thought, to the not too distant day when operations at their single runway aerodrome would cease and when only private flying would disturb the solitude of their picturesque and some-times quaint rural setting. The Essex farmers, who had awaited publication of the White Paper in a state of apprehension, could now look calmly across their fields of golden corn in the sure and certain knowledge that at least their environment was safe. They were, the White Paper had said, on the wrong side of London and too far from its centre to be a serious contender for the dubious privilege of having their sleep shattered by the screams of low flying jets, their thatched homes shaken, their winding hedgerowed lanes torn up to make way for the runways, hangars and scores of ancillary buildings that go to make a major airport. They had been saved, they thanked God, from the horrors of commercial aviation. Or so they thought.

For ten long and blissful years they lived in tranquil peace with the little-used airfield that bordered their turnips and their potatoes.

It had been built in 1942 by the United States Air Force as a war-time necessity. Good fertile fields had been lost then but this was part of the acceptable cost of war and everyone recognised that. Even after the war, when the Americans came back and extended the runway to 10,000 feet to make it one of the longest in Europe there were but few grumbles and no really serious opposition.

It was on a bright sunny evening in July 1964 that the bubble finally burst and the people of rural Stansted discovered how much confidence can be placed on a Government decision when the necessities of time become mother to the inventiveness of a political about turn. It was then, at a meeting at Harlow New Town, that local councillors, many of them representing midget authorities with few decision-making powers, were asked to give their blessing to the proposal that Stansted, which had been ruled out of the race more than ten years before, should now become the site for London's third international airport.

But it was not only the proposal itself which staggered them, but the sheer size of the plan. What was going to happen, they were told, was that 19 square miles of some of the richest and productive agricultural land in Britain was to be turned over to the "necessities" of jet age travel to become yet another concrete jungle base for the world's airlines.

The basis for the proposal, they were calmly told, was the report of an Interdepartmental Committee, whose findings had been published the previous March. The Committee had thoroughly examined, with great care, all the available sites for the now urgently required airport, and, after much heart searching and with due regard to the problems of the area, decided that Stansted was the only suitable place. Bewildered and shaken, the councillors returned home and eventually reported the details of the Harlow meeting to their local authority colleagues. There was immediate uproar and sometimes sheer disbelief that a site which had once been all wrong, was now all right. Parish Council and rural district clerks dug out the 1953 White Paper from their files, dusted it down and read it over again. Surely there must be some mistake. Stansted had not moved. It was still the same distance from London and still in the same place.

Slowly but surely, the full implications of the proposals dawned on the villagers. Not only were thousands of carefully tended acres going to be lost to agriculture, but 600 homes would have to be demolished and the entire hamlet of Mole Hill Green would disappear without trace. Furthermore, many more acres would be swallowed up as the airport developed and hotels, petrol stations and other service industries swelled the industrial infrastructure around the airport. The entire environment would disappear and be replaced by the ear-splitting, minute-by-minute screams of inter-continental jets as they cruised over the pleasant little towns of Bishop's Stortford and Great Dunmow which were more familiar with the sounds of sheep and cattle on market days and the distinctive East Anglian tone of the farmers in the pubs discussing stock prices.

The Committee which decided to inflict this burden on the

people of Stansted and district was headed by Mr George Vinger Hole, then Under Secretary in the Aerodromes' Planning Division of the Ministry of Aviation. One of the most astonishing things about it was its composition. Of its 15 members all but two had a professional interest in aviation. The others represented the Ministry of Housing and Local Government and the Ministry of Transport. Not a sign anywhere of a planning expert or anyone with any specialised knowledge of environment, particularly the problems of noise. The Hole Committee's terms of reference were simple—to inquire into a possible site for London's urgently-needed third airport. There were 18 meetings and each sitting reviewed a different site, they claimed afterwards. But strangely, they visited just one—Stansted. The Committee was born after The Select Committee on Estimates (attempting the yearly quart-into-a-pint-pot act of pruning Government costs and expenditure) demanded a complete reassessment of Stansted and its possible future usefulness.

Mr Hole, a dedicated believer in the progress of civil aviation, can be personally forgiven for his lack of expert knowledge on matters other than aeroplanes and airport planning. His job, and that of his Committee, was to quickly find a new airport to relieve the increasing congestion at Heathrow and Gatwick and one which, it later became clear, would overtake the pair of them in sheer size and operational capacity. While there is no reason to doubt the Committee's complete sincerity, it would be interesting to know which other sites they reviewed. How closely were the alternatives looked at?

However, their Stansted decision was backed by the Conservative Aviation Minister, Mr Julian Amery, who blessed it with the Government's official approval. This was in June 1963 and by fascinating coincidence the Wilson Report on Noise, published only a month later, indicated the increasingly grave concern felt in medical circles about the problems of noise as a direct environmental threat. Significantly, it was the Wilson Report which established a noise-measuring device known as the Noise and Number Index which is now an internationally-accepted measuring instrument for calculating the intensity and frequency of aircraft noise.

The Report said that a noise factor of between 50 and 60 NNI* in an urban area should be regarded as intolerable. In a rural area the same degree of intolerance was reached at a much

* Noise and Number Index.

lower level—45 NNI. It specifically highlighted the thousands of complaints from people living around Heathrow Airport where the Committee described the situation as "unprecedented". Wilson claimed that no less than 40,000 people who lived in the close vicinity of Heathrow were being subjected to more noise "than they can be reasonably expected to tolerate". It went even further and said the noise around that airport was the worst known in the country and pointed out that the people who suffered from it had no right of legal action to secure its abatement.

Wilson stressed it was vital that the noise lessons learned at Heathrow should be applied to other airports. In other words any future airport development should take into account noise problems that might arise if the development took place in a population centre. The Report said: "The important lesson is that as the noise extends far beyond the physical perimeter of the airport, the environs must be as carefully planned as the airport itself if serious noise nuisance is to be avoided."

One important point should be made here: the Hole Committee must have known that the Wilson Committee was sitting at the same time. Surely it would not have been beyond the bounds of ministerial protocol to find out the Wilson line of thinking on this most pertinent matter?

Had they done so, then a few more sites might have been scrutinised more closely and, when it eventually came to a Public Inquiry, the Government's case for Stansted may not have collapsed so completely as it did. Whether Mr Hole would have taken any notice of Wilson is a matter for debate. His attitude towards Wilson's recommendations, with all their concern for a little peace and quiet and the retention of some environmental stability, can be gathered from his reply to a petition from people living around Heathrow.

They were complaining bitterly about sleepless nights, how schools could not operate properly because of noise, and how their children could not rest because of roaring aircraft in the middle of the night. They wanted the noise to be reduced. Mr Hole dismissed their pleas as "impractical" and added: "Barring jet aircraft at nights is unacceptable and uneconomic."

So, as far as the Hole Committee was concerned, the Wilson Report recommendations were not worth the paper on which they had been written. Stansted was, in their view, the best site for the third airport and that was the only consideration worth talking about.

Nonetheless, the Committee—and indeed the Government—realised its proposals would meet very stiff and determined local opposition. But there had been opposition against Gatwick, too, and that had been successfully overcome. So why should Stansted be any different? There was great confidence in the corridors of Whitehall that the Essex villagers would work off their frustration and anger by writing futile letters to the Press and complaining ineffectively at a local Public Inquiry. Then, officialdom reasoned, the serious business of building a giant international airport would begin.

If the narrow recommendations of the Hole Committee was the first serious piece of Government miscalculation, then this attitude was certainly the second and became part of a long chain of blunders, misconceptions, half truths and, sometimes, absolute lies.

While, in Buckingham Palace Gate, engineers and planners of the newly-formed British Airports Authority were getting down to the detailed business of where the new terminals should be, and where the hangars should be located, the villagers around Stansted were sounding their call to arms. After the initial cries of anguish, village hall public meetings and angry, but largely uninformed, letters to the Press, the local residents knuckled down to the serious business of forming an articulate action group. They also started raising the necessary funds to thoroughly research their case so that they would be well represented at any Public Inquiry. And so was born the North West Essex and East Herts Preservation Association, a small group of Wellington-booted villagers who will go into history as the local yokels who took on the stiff-shirted might of the Government—and won!

One of the most impressive things about the Association's joint chairmen, Mr John Lukies, a magistrate and highly respected local councillor and farmer, and Sir Roger Hawkey, a businessman who lives in the area, was the quiet, unassuming and logical way they set about demolishing the Government's case for Stansted airport. They proved other alternatives were at least worth looking at and that some of these had the same, if not more, potential than Stansted.

Within only a few weeks came the first of a baffling and sometimes alarming series of contradictory statements: promises that, in the end, made it painfully clear even to the most unobservant onlooker that the airport's proposals had been almost unbelievably badly schemed, ham fistedly, amateurishly presented and this with a cynicism that clearly indicated all protest was useless. The arrival of the massive airliners at Stansted was merely a matter of the time it would take to construct an international airport.

On 28 August, only one month after the Harlow meeting the Ministry of Aviation denied the existing Stansted airfield area would be greatly expanded. Yes, it was true, a Ministry spokesman said, that a map showing a 19 square mile area had been openly and publicly displayed at the meeting but he shrugged this off as "only a rough guide". He blandly protested that even the runway alignments had not, at that early stage, been decided.

However, Mr Amery, the Aviation Minister, had already approved the recommendation and smilingly announced that work on the new terminal would begin within two years, by the middle of 1966. The left hand clearly did not know what the right was doing for, around the same time, Sir Alec Douglas Home, who was then Prime Minister, stated categorically that no decision on the airport had been taken.

Sir Alec gave this remarkable undertaking in a letter to Mr R. A. Butler (now Lord Butler), then Foreign Secretary and MP for Saffron Walden, one of the affected areas, who read it at a village protest meeting at Takeley which was perched on the perimeter of the proposed new airport. Mr Butler was obviously pleased with the assurances, and so were his constituents, who had been alarmed at Mr Amery's apparently dictatorial attitude. In his letter Sir Alec spoke of the impending Public Inquiry and said: "It is important that this Inquiry should be thorough and genuine, and I want to make it quite clear to you that the Government has in no way made its final decision and cannot do so until it receives the report of the Inquiry. . . . I further say that not only has the Government not made its final decision, but that the door is still not closed to suggestions or alternative solutions if they are put forward. This is, in fact, a long-term matter which can be decided only after a thorough Public Investigation has taken place."

If the Government was treating the building of a third international airport as a major planning consideration, it seems inexplicable that the Prime Minister would not know the attitude of his Aviation Minister. Yet it appears that he did not. Or was it that the Government and the Civil Service were treating the entire airport proposal as a foregone conclusion, and considered the odd conciliatory letter would in no way affect the issue and, in the long run, would be of little importance.

Or maybe the approaching General Election had something to do with it. Shortly after the Takeley meeting, Mr Butler was elevated to the Lords and Mr Peter Kirk, who very soon was to become a major figure in the battle over Stansted, was contesting the safe seat for the Tories. Not unnaturally, he took a great personal interest in the airport row and was one of the first people to realise just how flimsy and insecure were the foundations on which the Government had built its case. He received some temporary satisfaction when, one of his constituents received from Mr Harold Wilson, the Leader of the Opposition, a reply to a letter demanding assurances that a wide ranging, General Inquiry would be held into all the factors relating to the building of a major airport.

In view of the fact that he was soon to become the Prime Minister and that his own Cabinet would be seriously divided over the issue, Mr Wilson's reply is of significant importance. He said: "With regard to Stansted airport, the Labour Party is committed to a complete review of the internal transport system of the country . . . the question of air travel, including the siting of airports, clearly comes within the scope of our investigations and we would suspend the decision to develop Stansted until our planning is completed. We do not believe that the Government has paid sufficient attention to such important issues as the ever-increasing noise factor, and we would take this into account. We are also aware of the problem of travel between the centre of our cities and our airports, and are studying the possibilities of new and faster forms of travel. It may well be that this could enable the siting of new airports at greater distances from residential areas, while still ensuring faster transport arrangements . . ."

Mr Wilson may have been electioneering and so, for that matter may Sir Alec. But at least it appears that Mr Wilson had taken the trouble to read the noise report by his namesake and was astute enough to see its electoral possibilities. After all, apart from the loss of amenities and a man's own carefully tended fields, noise is one of the main objections to any airport development. After the Labour victory at the election, Mr Roy Jenkins became the new Minister of Aviation and was soon caught up in the complexities of the Stansted row. Not long after taking office he received a joint delegation from the Essex County Council—which violently opposed any airport development at Stansted

—and the Preservation Association. Shortly afterwards he sent the following letter:

"First it was represented to me that, unless the terms of reference (for the Inquiry) covered the question of timing, it would be possible for Ministers to make the entire Inquiry a formality by refusing to follow up indications in the Inspector's report that an alternative site was preferable to Stansted, on the pretext that the necessary study and survey of the alternative sites would take too long and the development of the airport could not wait. If the outcome of the Inquiry is that another site is to be preferred to Stansted, this will be followed up and it will not be ruled out for lack of time to study and survey the site."

In a letter to the Essex County Council, the Minister said: ". . . the deputation told me that a refusal to meet their request would tend to give the impression that the Government's mind was made up, that the Inquiry would be only a formality and that the development of Stansted would be pressed through regardless of the outcome. I can give you the firmest possible assurance on this . . ."

The Preservation Association committee had, in the meantime, been recruiting aviation specialists to its side and doing a great deal of homework on its own. One of their supporters was Sir John Elliott, former chairman of London Transport and director of B.E.A. Just before the Public Inquiry at Chelmsford, Sir John told a meeting of the Preservationists that, in his opinion, the plan for Stansted was "wrongly conceived and inadequately thought out". He went on: "Never in my forty years of public transport experience have I come across such inadequate and hurriedly put together plans."

All the careful homework was soon to pay off. The villagers had become so keen to win their fight and so enthusiastic that some of them were actually able to quote chunks of the 1953 White Paper, and the report of the Gatwick Public Inquiry verbatim. It was to be all good ammunition to fire at the ill-prepared Government when it came to giving evidence at the Public Inquiry.

This much-awaited, and soon to be famous, event began in the quiet of the County Hall of Chelmsford on 6 December 1965. The Inquiry was to sit for 31 days until it finally completed hearing all evidence on 11 February the following year and was to go down into history as one of the biggest defeats ever suffered by a Government at the hands of the public.

Mr G. D. Blake had been appointed as Inspector and Mr J. W. S. Brancker, his Technical Assessor. The Preservation Association, despite all the previous ministerial assurances, were angry at, but had to be content with, the Inquiry's terms of reference.

These were: "To hear and report on local objections relating to the suitability of the choice of Stansted for an airport and the effect of the proposed developments on local interest. It will be open to the objectors to suggest modifications to the outline scheme of development, or to propose alternatives, but not to question the need to provide a third major airport to serve London."

They argued that the terms of reference were too narrow, and should have considered the vital questions: "Is a third London Airport really necessary?" And: "when?"

From a sheer entertainment point of view, the Inquiry was to become one of the most popular events seen in Essex for years as, one by one, like lambs to the slaughter, Government witnesses went meekly into the witness box to have their evidence torn to shreds before the eyes of Press and public. It had been a long time since the men of Whitehall suffered such public humiliation. The blow to their pride and dignity must have been fearful, like being caught cheating at bridge in the club.

There was a sensation straight away when a thorough report, carefully prepared by Messrs Norman and Dawbarn on the complex engineering problems involved in developing Stansted, showed that the interdepartmental committee's proposals were changed even before the Inquiry began! The new proposals, they reported, were designed for two parallel runways, one capable of reaching 16,500 feet, the other 12,000 feet and the pair separated by not less than 6,000 feet. The design was for a standard aircraft movement rate of 64 an hour with a peak of about 80 an hour.

The Ministry of Aviation's case was presented, to the very best of his brilliant ability, by the late Sir Milner Holland, Q.C. But, one is led to believe subsequently, the instructions that he received may well not have been entirely to his satisfaction. A punctilious man ever aiming at perfection, his impatience with the imperfect became evident when he allegedly remarked wistfully to a colleague: "I must say, I have never been so badly or inadequately briefed in my entire career."

It became all too clear from the start that Sir Milner was batting on a very sticky wicket. He was in the frustrating position of having to stand back helplessly as, one by one, his expert witnesses were decimated by the withering cross-examination of Mr Peter Boydell, Q.C., for the Preservation Association, and Mr Douglas Frank, Q.C., for the Essex County Council.

The Inquiry encompassed 11 long and tiring weeks and around 200 witnesses were called. More than once, Sir Milner left the County Hall after a day's hearing shaking his head in utter disbelief at the impossible task he had been set. Any lawyer knows that without the right ammunition you cannot win a legal war, and Sir Milner just did not have such ammunition. The theory even arose that the authorities had already decided that the airport would go to Stansted, no matter what anyone else said. Even so, the most pessimistic Government official did not image for one moment that the case for the airport would take such a heavy pounding.

Sir Milner opened with a mighty and impressive speech, immediately repudiating suggestions that a Government decision to build London's third airport at Stansted had already been taken. "Full weight" would be given to the Inspector's findings and observations. Before any airport was built for London, it must satisfy five main requirements:

1. compatibility with Heathrow and Gatwick, so the three could operate together from the point of view of air traffic control;
2. air safety;
3. convenience for air routing;
4. easy reach of London;
5. Land suitable for airport construction.

Furthermore, Sir Milner added, the third airport must be within a maximum of one hour's travelling time from Grosvenor Square.

If the Whitehall officials had bothered to show Sir Milner the 1953 White Paper and quote him the relevant paragraph, he would immediately have seen that one of the reasons against Stansted at that time was that it was up to two hours travelling distance from central London. It was a curious error, and the Preservationists, sitting at the back of the room, gleefully made note.

Furthermore, he said, any airport development had to satisfy six "negative requirements". These were:

1. reasonable noise nuisance;
2. compatibility with regional planning;
3. compatibility with agriculture;
4. compatibility with national defence;
5. no unreasonable conflict with civil aerodromes and due consideration to the requirements of private flying and gliding.

Sir Milner then made one of those alarmingly categoric statements that were regretted afterwards and were thrown back in the Government's face time and time again. For he gave an absolutely firm undertaking that unless all these requirements were satisfied it would be impossible to proceed with the third London airport and "that would be the end of the matter".

In view of what has happened since—and indeed what happened at the Inquiry—it is not surprising that the baffled Sir Milner considered his task was well nigh impossible. He must have been very close to tears of anger and frustration on the day that Mr Hole gave evidence about the future of the United States Air Force base at Wethersfield, only a few miles from the proposed airport site. Mr Hole said it was obvious that if Stansted was built the U.S.A.F. base—parts of the N.A.T.O. command of the Nuclear Strike Force of the 20th Tactical Fighter Bomber Wing—would have to close down, and its closure would, of course, cost many millions of pounds. Under close cross-examination about the base's future Mr Hole surprisingly announced to a staggered audience that a decision to close the Bomber station had already been taken. He rocked everyone, including Sir Milner Holland, when he calmly announced that the decision had been taken since the Inquiry opened!

Mr Boydell swooped on him like an alert eagle on an unsuspecting field mouse. He was most surprised at Mr Hole's statement because inquiries had been made that very morning with the head of the United States Air Force in Britain. And not only was the American Air Force unaware of any such plans for closure but, were in fact going ahead with costly improvements, including the provision of married quarters and more long-term administrative buildings.

It was a grave blow to the Government case. If Wethersfield was not to close, then surely London's third airport could not be built at Stansted, reasoned the now smiling Preservationists.

Very soon, Mr Hole was forced to concede yet another major point. Gloomily, he agreed that Government policy dictated that valuable agricultural land should not be developed unnecessarily. He nodded further reluctant approval to Mr Boydell's suggestion that not one agricultural expert had served on his committee and that, furthermore, the committee had not even considered either the value of the land being lost, or the value of its produce.

The official case was now obviously going very badly, and was not helped at all when Mr Andrew Miller Kerr, Assistant Chief Engineer in charge of motorways at the Ministry of Transport, admitted under cross-examination, that even with the new dual three-lane motorway access, it would take between 85 and 95 minutes to Stansted even at off-peak periods. This was a shattering admission and it completely undermined one of the most important of Sir Milner Holland's original conditions—that any airport should be within "a maximum" of one hour from Grosvenor Square.

Mr Boydell was naturally delighted. He even went so far as to propose that his clients had no case to answer. "We are saying that this case has been gravely ill-prepared and it is shaky to say the least. The Minister must think again as far as Stansted is concerned," he told the Inspector.

This was the pattern of events throughout most of the Inquiry. The Government case was destroyed at every turn. It was now more than abundantly clear that not only had the Hole Committee done insufficient homework, but that most other Government departments were in a similar position. It was hard not to draw the inference, once again, that the Government's intention was to steamroller the airport opposition with the minimum of legal difficulty and start construction as soon as it became politically feasible. The Preservation Association and the Essex County Council were delighted at the way things had gone. The Preservationists had raised £25,000 from more than 13,000 subscribers in order to present their powerful case, and they felt it had been worth every penny.

Furthermore, the Inspector was left in absolutely no doubt at all what effect a massive international airport would have on the amenities in this delightful part of rural Essex. Several witnesses spoke of the area in almost reverent tones. They painted verbal pictures of quiet agriculture and lovely old Saxon and Norman churches, high hedged winding little lanes, skilfully thatched farms and cottages, and graceful old Tudor manor houses surrounded by spreading green fields, dotted with old elm and oak.

This was compared with the sprawling urban development which would follow in the wake of any new airport. The Inspector was reminded of Heathrow and the hotch potch of hotels, motels, warehouses, long sprawling maintenance areas, the inevitable "little box" housing estates to accommodate the expected 30,000 employees and their families who would be drafted into the area. It was an uninviting contrast and that was just what the Preservationists intended. Sensing the Inspector's anger at the badly-prepared Government case they decided to make the best of it. They piled tale upon sorry tale of how a hospital serving a wide area around Bishop's Stortford would have to close, and pointed to the enormous expense of sound-proofing, or indeed replacing, fifty schools to prevent them being subjected to such tremendous jet noise that teaching would be impossible. The cost: the locals emphasised the cost, the cost, the cost! The Government had made no mention of this and had not taken it into account. Officialdom was looking dejected. Certainly, it had not expected to be bombarded in such a well-informed, technical manner by a group of villagers.

In view of what was to happen later, Mr John Brancker, the Inquiry's technical assessor and an aviation planner of international repute, made what turned out to be an almost prophetic observation. He thought other sites could have been considered with a greater degree of seriousness if the Shoeburyness artillery range—the greatest obstacle to the development of Foulness and, to a certain extent, of Sheppey—was moved elsewhere. The military was having none of it. They proved their case by sending along Major General Egerton, Director General of Artillery, who categorically stated that a move to another range was out of the question. There was not one available.

Mr Brancker was not convinced. A base so near to London was an anachronism and consideration should be given to moving it, he said.

It was clear that Mr Brancker, the Government's own choice as an independent assessor, was not satisfied that Stansted was the only possible site available for a third airport. Without wishing to encourage delays he explained that "much of the evidence submitted seems to me rather superficial". A wide-ranging examination of the situation before any decision was taken would make him feel very much happier.

And so, in February the Inquiry closed. The technical experts went away licking their wounds and the Preservationists went back to ploughing their fields and milking their cows, confident that if there was any justice left in the world of ministerial decisions victory would be theirs. It had been made abundantly clear to the Inspector that other sites had not been given more than fleeting consideration. Luckily, they kept their propaganda machinery in good working order, for soon their dreams of justice were to be destroyed by an under-the-counter political move that sent even the most ardent supporters of the Labour Government creeping to quiet corners blushing with shame.

Three months later Mr Blake, the Inspector, presented his report to the Minister of Housing and Local Government. The Minister, reading it in the pivacy of his office in Whitehall, must have been faced with an unenviable dilemma. For one paragraph stood out as brightly as landing lights on a runway. It said: "It would be a calamity for the neighbourhood if a major airport were placed at Stansted. Such a decision could only be justified by national necessity. Necessity was not proved by evidence at this Inquiry."

According to Mr Blake's report, the official proposal succeeded only on the viability of air traffic. On five other counts the scheme failed. No evidence, he said, had been produced that Stansted was the right place for the "traffic focus" a major international airport would bring. The evidence, in fact, had been to the contrary. Therefore it failed from the point of view of town and country planning.

Road access to Stansted from London was bad, and the proposals would be unacceptable to passengers and airlines "to an extent that might make the airport of only moderate value". From the point of view of noise, Mr Blake believed that restrictions he thought necessary to impose "would materially restrict capacity operation". The change in the character of the neighbourhood would, he said, cause great local resentment, mainly because of noise and traffic nuisance. Finally, the loss of "many thousands of acres" of good agricultural land also weighed against the success of the proposal.

The Inspector bludgeoned the Hole Committee by noting: "In my opinion a review of the whole problem should be undertaken by a committee equally interested in traffic in the air, traffic on the ground, regional planning and national planning. The review should cover military as well as civil aviation."

The report caused not only considerable consternation in Whitehall but clearly shocked the Government who saw their third airport runways disappearing in neat rows of cabbages and runner beans, and Mr Blake was not the most popular man around Whitehall. The Government had expected, in view of the way things had gone at Chelmsford, to have their wings cut back a little. But such a damning report was certainly not expected and their confidence in the Stansted proposals being implemented quickly began to wane. It was obvious that the bulldozers and heavy construction equipment were not going to lumber over the fields of Stansted for some time to come.

But what of the politicians? They too had expected the Public Inquiry to be a routine affair where the prisoner "after a full and fair trial will be found guilty and sentenced to death". But the prisoner had been found not guilty. Surely, by all the rules of legal procedure, he should now be freed and after his harrowing experience sent home to pick up the threads of his interrupted life.

But the politicians needed a third airport—and quickly. What is more, they were determined, no matter what the Inspector said, that the airport would be at Stansted. Politics being the art of the possible, they sought a solution which would give the B.A.A. its airport and at the same time satisfy the people of Stansted that they were getting a fair deal.

The Inspector wanted a review of the whole problem. Very well, he would get one. But it turned out that the "review" could be more accurately described as a retrial.

Furthermore, it was a retrial held *in camera* with the prisoner not even knowing the decision of the earlier hearing. For Mr Blake's report was not made public for almost a year and the composition of the review body was never a matter of public knowledge. To this day, its members have never been identified. There are even those around Stansted who believe emphatically that it was the Hole Committee in disguise. Certainly, it is curious that a Government body, examining something as vital to the national interest as a giant international airport, should conduct its business in secret. Even more alarming to anyone interested in the activities of a democratic society is that the objectors were never given a hearing. How could they be? They did not know a review was taking place or what Mr Blake had said in his report.

They only found out on 12 May 1967 when Mr Blake's long-awaited verdict was published simultaneously with yet another

White Paper—"The Third London Airport"—which was said to be the result of the review body's inquiries. To publish them together was a blatant political confidence trick aimed at presenting Parliament, the Stansted objectors and the whole country with a *fait accompli*. What is more, the deft sleight of hand manoeuvre almost succeeded.

What happened was this: the responsibility for airports and aviation planning had switched from the Housing Ministry to the Board of Trade, whose President was Mr Douglas Jay, allegedly dedicated to the idea of the third airport being built at Stansted. When he received the Inspector's report of the Public Inquiry, he was in a difficult situation. He had been told by Mr Peter Masefield, Chairman of the British Airports Authority, and other experts, that a third airport must be operational by 1976, otherwise valuable dollar-earning aircraft would have to land at continental airports, like Amsterdam and Paris, because Heathrow and Gatwick would be too busy to take any more planes. Clearly that situation must not be allowed to arise. But what was he to do? The only airport likely to be ready by the B.A.A.'s middle seventies deadline was Stansted, and here was a Government-appointed Inspector saying, in effect, that the site was useless, that the Hole Committee had made a blunder, and that the situation ought to be thoroughly re-examined.

Mr Jay made two political miscalculations which eventually cost him dearly and may, in the end, have been partly responsible for his return to the back benches. Firstly, he appointed the secret review body and refused to disclose the names of its members, or how many times it met, or which other sites, apart from Stansted, they had closely examined. Secondly, he published Mr Blake's report and the White Paper on the same day. Everyone was given a classic opportunity of comparing them side by side—and the all too obvious indeed glaring inconsistencies which existed between the two documents naturally gave rise to rather serious disquiet. The White Paper said that after a further re-examination of the problem, the Government was of the opinion that Stansted was the right place for the third airport, and that was where it was going to be built.

The timing of the simultaneous publication also aroused speculation of political jiggery-pokery. Friday, 12 May. Mr Jay must have known that the majority of MPs would, by then, already have left London to start their holiday. The House was nearly empty when the White Paper was published, and he was ensured a clear three weeks free of parliamentary criticism. Enough time perhaps for outraged tempers to cool and for him to rally more support to the drooping Government flag. But Mr Jay was already in serious trouble with some members of the Labour Cabinet, some of them objecting to his strange handling of the situation, and also remembering the promises and assurances they had given to the Preservation Association and the Essex County Council. Clearly the Government was in a cleft stick. Nobody doubted that London needed a third airport, but several very senior politicians were doubtful whether Stansted was the place for it.

There were a number of rows at Cabinet meetings and a wide measure of disagreement about what should be done, particularly when it was discovered that a large number of Labour MPs had signed a petition demanding a re-examination in public of the whole problem.

Eventually, after a long and rowdy debate in the Commons, the Government had to slap a three-line Whip on Labour MPs to make sure they voted in favour of the White Paper—a startling state of affairs over something as "ordinary" as a planning decision. But, in the event, it was very necessary. A free vote would almost certainly have meant a defeat for the Government which—although it probably would not have fallen—would at least have had to abandon any hope of building the airport at Stansted.

There is one significant and brave exception. Mr Stanley Newens, then Labour MP for Epping, defied the Whip and sat steadfastly in his seat and did not vote at all. He, at least, had the courage of his convictions. Throughout the long battle to save Stansted, Mr Newens figured prominently and often found himself in political hot water because of his avowed opposition to the airport scheme. He was many times threatened with dire punishment if he did not keep quiet. This he flatly refused to do and he and Mr Peter Kirk, the member for Saffron Walden, formed a sort of unholy alliance and, remarkably, found themselves on the same side in the irregular Commons rows about the airport scheme.

When the Public Inquiry report and the White Paper were published, the Preservationists went screaming in anger to their nearest meeting place. They were particularly incensed because they had made it clear they would be willing to accept Stansted as the third London airport—although many of them had plenty to lose if it came—"so long as the choice was publicly shown to be the best one possible". Now they were absolutely certain that Mr Jay, and the entire Government, had attempted a parliamentary manoeuvre that would make the development of Stansted airport inevitable.

With the White Paper safely steered through the Commons it looked as though the fight to preserve Stansted was over. Most protest groups would have given up all fighting and surrendered in sheer frustration on the basis that you can't beat the system. But Mr Lukies, Sir Roger Hawkey and their fellow Preservationists were not through yet. They appointed a Press Officer—Mr Lukies' 24-year-old daughter Susan—and sent her to proclaim their Public Inquiry victory and protest loudly at the "iniquitous and underhand trick" that had been pulled by the Government to achieve its ambitions. She was told to prove that the Inquiry had been a farce and that the Government's claim to impartiality was little more than a mocking charade.

It would have been a big task for any high-powered public relations outfit. For a young married woman, who had never been to Fleet Street and knew not a single journalist, it appeared almost madness.

CHAPTER THREE

The Day Sue Came to Town

Public Relations Officers tend to come into two grades—the very good and the very bad. The good ones are charming, friendly, helpful and informed. The bad ones are charming, friendly, obstructive and clueless. Both types claim to be experts in their particular subjects and can usually be found in Fleet Street hostelries drinking with their newspaper "contacts". In many cases, the professional PRO is himself a former journalist and therefore experienced in knowing the "angles" which, he hopes, will ensure publicity for his product in either newspaper or magazines.

To the Fleet Street veteran, the seasoned, tough reporter, "the Street" has a sort of presence, a tradition of Johnson and Boswell, of Swaffer and A. P. Herbert, of Edgar Wallace and Charles Dickens and Chesterton. It is a place where the newcomer has to prove himself, where dreams are built but rarely become reality; where hopes are raised only to become empty illusions. It has variously been described as the Street of Adventure and the Street of Misadventure. It often champions apparently lost causes and sometimes even wins. Fleet Street, to many, is the place where injustice is occasionally put right and where miracles happen intermittently.

And it was a little miracle that young Mrs Susan Forsyth, then aged 24, was looking for when she left her thatched cottage home at Henham, Essex, early one Monday morning in the summer of 1966. She had just been appointed Press and Public Relations Officer for the Stansted Preservationists at the princely salary of £600 a year.

To many Stansted was already a lost cause. The Government had made its irrevocable decision. And all but the supreme optimists among the Preservationists were losing heart.

In Fleet Street many journalists who reported the fight to save Stansted were convinced that the airport was as good as built. They saw the injustice, but they had seen that before, and steamroller Government was nothing new to the experienced reporter. Although newspapers expressed disgust at the way the Stansted affair had been handled, it now seemed that further effort was likely to be wasted.

But young Susan was eternally optimistic, perhaps because of her naïvety of the inner workings of Whitehall and the machinations of the national Press.

However, when she parked her green Mini that morning and stepped for the first time into the rarified atmosphere that is Fleet Street, she not only did not have the first idea of how or where to meet a journalist, but she had never met one before in her life. At that stage they were merely names underneath headlines, ghostly figures who composed breakfast table reading.

But, when her job was completed almost three years later, the Roskill commission had excluded Stansted from its short list of four sites and Susan had been partly responsible for a Government Minister leaving the government. She sparked off newspaper campaigns that ran into millions of words, a Commission that cost £1,200,000, and, most important, helped in the salvation of one of Britain's most beautiful rural areas from the commercial necessities of jet age aviation.

By then she was a popular and well-known figure among many Fleet Street journalists, and the envy of scores of highly paid professional PROs who would, with some justification, have thought their charges modest if they had conducted a similar campaign for the Preservationists and sent them a bill for £100,000. Susan cost them less than £1,800.

The first thing she did was to contact some of those anonymous newspaper by-lines by telephone. It was a simple approach. "Oh, my name is Sue Forsyth and I wonder whether I could talk to you about Stansted airport? . . . Yes that's right, I am from the Preservation Association. We think we have had a raw deal and I would like to talk to you about it. We still believe that we are right and we have a campaign that might pull us through. Do you think I could meet you somewhere and we could chat it over."

She was mildly surprised when some journalists made excuses and others would not see her anywhere but over a drink in one of the Fleet Street pubs. That, almost certainly, was not the way Susan had envisaged her business would be transacted.

But, she soon discovered, that was Fleet Street's method and she realised that very few journalists will talk to you in church.

While Susan was perplexed by the insistence on public house conferences, the journalists themselves were equally intrigued by the attractive voice on the other end of the telephone, and mildly curious about what message she could possibly have that had not already been said before. Anyway, what could she tell them that they did not already know? When they kept their saloon bar rendezvous, reporters were surprised to find a five-foot-four brunette, neatly dressed, who blushed easily and certainly was not used to the sometimes straight-from-the-shoulder language of Fleet Street's bars.

One reason for the reluctance of some journalists to meet her was that reporters have an in-built antipathy towards public relations officers as a breed, and those who plead for publicity in particular. The females of that profession can sometimes be hard and unfeminine. But in Susan most of them found a quiet young woman who was intelligent, charming and as eager to listen as she was to talk. She said afterwards that those first few weeks in Fleet Street were nerve-wracking and often tearful experiences. She arrived with a simple message—that Stansted was wrong and that surely everyone must be able to see that. She soon realised that it was not that easy to convince the journalists of Fleet Street, particularly those who knew she was not a lone voice bleating in the wilderness. Behind her was a very powerful and well-organised Preservation Association which had, as it turned out, sources of information both near to and inside the Government itself. Among their sympathisers, for instance, were several members of the House of Lords, more than a handful of MPs, the entire Essex and Hertfordshire county councils, the National Farmers Union, the Noise Abatement Society, the Royal Institute of British Architects and the council for the Preservation of Rural England. Backing them were more than 70 affiliated societies.

None the less she was the voice for many of these people in Fleet Street and it was up to her to convince newspapers and magazines that the fight against Stansted was a justifiable one and that there had been a gross miscarriage of justice.

Apart from the long and sometimes wearying business of chatting up journalists, one of Susan's main occupations was reading. Every morning she scoured the Press to find any reference to Stansted or, indeed, to any airport proposal throughout the world. She also took a particular interest in the parliamentary reports and made careful note of every question and answer relating to aviation in general and to Stansted in particular. Hansard became her daily Bible. She made a careful and laborious study of the highly technical aviation press. The result of all this reading was the collation of propaganda ammunition resulting in 31 packed scrap books of cuttings from scores of newspapers and magazines, in addition to some of the rather more obscure journals.

She was so meticulous that she was often able to give a detailed off-the-cuff answer to statements, made primarily by Mr Douglas Jay and Mr Peter Masefield—often proving they had contradicted earlier statements. She compiled a notable list of these conflicting statements which she used as ammunition against the two officials to show how inconsistent they were.

As the weeks slipped by more and more journalists became familiar with the telephone number at Great Dunmow in the small backroom office in a gracious Tudor home that Susan used when she was not badgering the men of Fleet Street.

The house, Greencrofts, belonged to retired businessman Mr Edward Judson and his wife Eileen, two of the Preservation Association's committee. As Mrs Judson said when the campaign was at its height: "We used to live a nice quiet existence here. Now the phone never stops ringing day or night." Visiting journalists were always welcome and usually given a handsome glass of scotch or a glass of beer. But it was the information they were after. The latest "state of play", the latest background.

The small backroom began to look more and more like an efficient press office. The newspaper and magazine column inches began mounting up and the supposedly lost cause of Stansted began to revive. Probing journalists, now genuinely believing in the chances of the campaign's success, themselves began "digging up" all sorts of relevant material which put the Government in general, and Mr Douglas Jay in particular, in a more and more embarrassing position. The President of the Board of Trade attempting to justify his rigid attitude, only succeeded in contradicting himself more and more. And this, in turn, provided Susan with an increasing amount of ammunition to fire back at him via her new-found contacts in Fleet Street.

It was around this time that interest began to focus on Foulness as an alternative possible site to Stansted. Although, in the past, the little marshy island on the Essex coast had been considered as a possible airport site, the new proposals caught the public interest. The plan envisaged reclamation of large areas of the Maplin Sands which would be built on. The island would remain untouched. The idea of pushing back the sea to provide more land was not new in itself. Everyone knew the Dutch had been doing it very successfully for years. But, for some reason, it fired public imagination in Britain. The national Press was intrigued by the possibilities and began investigating the potential with feature and background articles, including illustrations of what could be done by showing what was happening at Rotterdam's Europoort, and the large redevelopment schemes at Antwerp, Marseilles, Dunkirk and Le Havre.

Air correspondents were also quick to point out some of the other advantages of a coastal airport because, during that summer, two jets crash-landed in the sea alongside the runway at Kai Tak in Hong Kong. In both crashes only one person died. What would have happened, asked the Press, if those jets had come down around busy Heathrow, Gatwick or Stansted. The casualty figures must have been much greater, and at the worst everyone aboard both aeroplanes may well have died. It was a powerful argument in favour of a coastal site airport.

The idea of building Foulness on reclaimed land was the brainchild of consulting engineer Mr Bernard L. Clark of Victoria Street, Westminster. The timing of its arrival could not have been better if it had been planned, because national newspapers running full scale campaigns against the Government's decision had been running short of new ideas.

Foulness provided a whole new field of thought and activity, not only for the Press, but for Susan Forsyth as well.

Apart from the written word, Susan, her father, Sir Roger Hawkey and other members of the committee had harnessed television. One of their neighbours at Stansted was Mr Aubrey Buxton, chairman of Anglia Television, who they interested in doing a documentary programme about the possibilities of a Foulness airport.

Not to be outdone, the BBC television people weighed in with discussions, news broadcasts, and documentary features. The real significance of this was that the emphasis of the campaign switched from the negative aspect of merely opposing the Stansted proposals to the positive aspect of supporting a new venture. In no time at all, the Foulness band wagon was rolling merrily. Everyone was suddenly enthusiastic about the idea and this was heightened further when the plan to reclaim land for an airport was extended to include a massive deep water seaport, similar in size and concept to that being built just across the North Sea in Rotterdam.

Young Susan's brief was now considerably expanded. Now she was not only opposing Stansted but proposing Foulness and putting up a case of considerable strength and depth in favour of the coastal site which, she never failed to point out, had never been seriously examined by any Government because of what was thought to be the immovable object of the Shoeburyness firing range.

With few exceptions, the national Press moved more and more in favour of a Foulness development. They recognised the prospect of developing a jet age airport, with built-in air safety advantages because of its location and saw the surrounding industrial infrastructure as an economically viable proposition. One of the immediate results was that previously disinterested public opinion now began to appreciate the advantages of a coastal site, especially as it would avert the tragic crippling of a thriving and prosperous rural community.

The influence was felt in the House of Lords, where a power group of forty peers formed their own pro-Foulness committee, were lectured by the people backing the coastal scheme, and even at one time suggested bringing in their own private Foulness Bill.

On her almost daily visits to Fleet Street, Susan's calm confidence reflected a growing optimism among the villagers of Stansted, the now veteran campaigners.

They clearly felt there was a chance of victory after all. At the end of the dark tunnel there was a small chink of light. It also became known in "informed circles" that many senior Government planners were realising that a re-appraisal of the Stansted situation would have to take place.

The method, thoroughness and persistence of Susan's Fleet Street campaign was later to provide an action blueprint for protesters who objected when Mr Justice Roskill's commission drew up its list of four possible sites. Indeed, the action group at Nuthampstead pleaded with her to conduct their campaign for them. This she did, but mainly in an advisory capacity. She was in close touch with the Nuthampstead villagers until she discovered, in early 1970, that she was to become a mother. Her daughter, Jacqueline, was born in September, 1970, but Susan is still keeping a very close eye on the battle for the third airport. "I think that the Preservation Association's campaign has proved one thing: if you know you are right and feel you are right the chances are you will win in the end. But I must say there were occasions when I thought we would never have made it.

"Now it has all been worthwhile. At least Stansted has been saved for the time being. What happens in the future, I just do not know. After all, there is talk already of a fourth airport. I just hope we don't have to go through it all over again."

CHAPTER FOUR

Brancker's Black Book

Mr John William Sefton Brancker is a monocled sixty year-old. Immaculately dressed, utterly precise in manner, he drinks large gins, smokes many cigarettes, and is prepared to talk quietly but endlessly about airports. His knowledge of the subject goes back to 1929 when most of the world's airports were little more than primitive landing strips, air traffic control was almost unheard of, and jets were not even beyond the preliminary design stage.

It was then that Mr Brancker joined Imperial Airways. Since then he has been variously manager and director of several African airlines, BOAC's Regional Director, and Deputy Assistant Director of the International Air Transport Association—the world "parliament" for 106 airlines. On the way, he picked up an international reputation for solving the technical problems and the mass of complex requirements that are needed to make any airport economically viable and operationally safe.

It was this international standing that led the Labour government to appoint Mr Brancker the independent technical expert at the important Stansted Airport public inquiry. His job was to assist Mr Blake, the inspector, on matters of airport technology. His aviation pedigree and acknowledged integrity made him a natural choice for the job. Throughout the inquiry he sat alongside Mr Blake and listened intently to all the evidence from both sides. Later, the task completed, he returned to his home in Montreal, having already established, through his close contacts with Mr Blake, what the inspector's views of Stansted were.

Then, more than a year later—on 12 May 1967—Mr. Douglas Jay, President of the Board of Trade, published Mr Blake's report. At the same time he issued the government White Paper which outlined the government's proposed decision to build the airport at Stansted—despite the welter of adverse evidence produced at the public inquiry and the criticisms made in the inspector's report.

Mr Brancker, normally a quiet, even-tempered man, was so incensed at the decision that he was moved to write a surprising letter to *The Times*. He did this, fully realising that Mr Blake, as a civil servant, could not publicly air his views on what the government had done and the interpretation it had put on his report. No such silencing restrictions applied to Mr Brancker.

He was a completely independent hired man, and, therefore, not subject to the same rules of official silence.

The letter itself was of significant importance. Not only did it show Mr Brancker's distress at the government's decision. It again brought into public question the credibility of an administration which had glibly claimed to have examined all alternatives to Stansted very carefully, and to have "reviewed" the situation in accordance with the recommendations in Mr Blake's report.

In his letter, published on 29 July 1967, Mr Brancker pointed out that, from the point of view of air traffic control, Stansted was quite satisfactory. Of the sites submitted to the Inquiry, Stansted was the best—assuming the permanent existence of the Shoeburyness Firing Range. But if that rather absurd anachronism could be removed, the advantages of Stansted would be more marginal. The key to the correctness or otherwise of Stansted lay in another direction.

The building of a major airport was equivalent to building a seaport. Any Government should therefore be prepared to accept the inevitable surrounding development that would follow. This included the provision of all modes of transport.

"In spite of being in the business for more than 35 years, I do not believe that air transport has the right to inconvenience the public at large without limitation: it exists to serve people, not to annoy them," Mr Brancker explained. But he added a warning:

"Certain economic necessities must be recognised if the airlines and the airport are to operate efficiently. There must be no restriction . . . on the use of runways or the number of landings and departures of normal transport aircraft because of noise, because such limitations inhibit the proper use of very expensive capital assets.

"Let us remember that, because of bad planning, both Heathrow and Gatwick are already restricted on this account, and the position in this respect is in no way improved by leaving the noise problem to be solved by the British Airports Authority, which is what the White Paper suggests.

"I have consistently maintained that airports and residential areas do not go together."

What he did suggest was that high noise areas surrounding airports should be either left as open spaces, or become part of an industrial complex with sound-proofed buildings.

But the real sting came in the tail: "If the Government is ready to accept this approach to Stansted, without recrimination when further land is required, without limitations on the use of the airport in terms of the times and the number of movements, and with the recognition that more expenditure will almost certainly be needed to provide adequate access, then there can be no further objection from my point of view.

"If these conditions cannot be met on a long-term basis, for political or other reasons, then the choice is the wrong one because an interminable clash between the economic necessities of the airport and a contradictory social environment will prove very unfortunate for all concerned . . ."

The Stansted Preservationists were delighted. If the Government's own technical expert was unhappy at the situation, he could prove a very powerful force on their side. After all, Mr Brancker was a potent piece of ammunition.

They wrote to him at his Montreal home and asked whether he would write two papers on their behalf. The first, an analysis of the much-criticised White Paper; the second, his views on a national airports policy. Partly because of his genuine sympathy for the plight of the Stansted people, partly because of his anger at the Government, he agreed.

So, after the official White Paper came the "Stansted Black Book", which incorporated the two requested papers.

CHAPTER FIVE

White Paper Attacked

Mr Brancker's look at the White Paper was soberly described as "an analysis". Never has a document been so misnamed. It was a lethal weapon, a crushing bombardment of the Government's case for developing Stansted.

At the same time, he threw some deadly high explosive at the report of the Inter-Departmental Committee which had proposed Stansted in the first place. He found large gaps in the Whitehall defences and exploited them with the tactical skill of a brilliant general who has suddenly discovered all the opposition's tanks were made of cardboard.

Firstly, he took an enormous swipe at the Inter-Departmental Committee report which had decided that the countryside around Stansted was "suitable for airport development". Mr Brancker came back: "This may be so in the civil engineering sense, but not in any other." In other words, apart from the obvious fact that it was physically possible to build a major airport in the middle of the Essex countryside, the proposal fell flat when items like environment, amenity, access, and so on were considered.

The same committee also said: "Parallel North-East/South-West runways should cause no serious noise problem." This was a gift for the expert who had turned against the Government:

"While this has not been repeated in the White Paper, it makes one somewhat doubtful of the validity of any of the other statements in the Inter-Departmental Report—including those ruling out the suitability of other alternatives—because the evidence of the (Public) Inquiry showed very plainly that quite serious noise problems will arise."

Not bad for a start. But things were warming up. The stripe-trousered Whitehall planners were beginning to squirm under their bowler hats. The White Paper had tried to prove that a third airport was vital by the middle of the 1970's, Mr Brancker continued. However . . . "during the Inquiry itself, virtually no evidence was offered by the Ministry of Aviation in support of its own traffic forecast."

Furthermore, little information had been given to the Inquiry about the origin and destination of people using London's airports. It was only later—in the White Paper—that the results of a traffic survey were made known. These showed that 80 per cent of London's international traffic was generated in the metropolitan area and the South-East.

"One must ask, however, why these figures were not available before the Inquiry. If they had been . . . the position would have been much clearer. The surveys do support the need for a third airport, but the decision that one was necessary was taken by the Government before the figures were available."

Even these survey figures, helpful though they were, still did not indicate the actual points of origin and destination of passengers within greater London—a necessary piece of information when making a choice.

In short, Mr Brancker was saying that it was no use putting the airport in the middle of Essex if the bulk of passengers came from Surrey. The traffic forecasts also suffered from another "very serious omission"—the effect on runway capacity of the growth of freight cargo.

The White Paper, he recalled, had estimated that 43,600,000 passengers would be using Heathrow, Gatwick and Stansted by 1980. But there had been no mention of cargo growth. "And, considering the fact that this form of traffic is growing extremely rapidly on a world-wide basis, this . . . throws some doubt on the validity of other forecasts."

There had been a suggestion that any third airport could be used solely as a short-haul terminal and this was "worthy of scrutiny". In other words, a sort of commuter airport which would deal exclusively with inter-Britain flights, the quick forty-minute hop between London and Manchester.

Although there were obvious objections in a place like London, there was nothing new in the principle. La Guardia had been doing the job effectively for New York for some years and the National Airport the same for Washington. To some extent, Southend already played a similar role for London. The objections were valid, but the White Paper's reasons were "not wholly convincing".

"We know already that short-haul operations form a considerable percentage of aircraft movements and the growth of services of this kind might be concentrated in one spot," he contended.

One cannot help feeling that Mr Brancker, at least occasionally, regarded the logic of the argument in the White Paper rather as a Don would review a theological thesis by a below-average first-former. This impression was particularly clear in the following assessment:

"It is also stated in the White Paper that the inevitable moves from Heathrow to one of the other airports by foreign, as well as British airlines operating short-haul services, would be unpopular . . . and that retaliation must be expected against our national airlines in foreign countries.

"If the short-haul airport was inconvenient and inaccessible, this would be true. But if the foreign carriers found it more accessible than a long-haul alternative—and if it reduced his air mileage—he would probably be delighted to use it.

"The danger of retaliation would arise if he were compelled to use a third airport which had less passenger-appeal than one used by his British competitors. The White Paper itself states that the third airport, wherever it is sited, cannot be as accessible to central London as Heathrow."

After further attacks on the White Paper's line of reasoning, its hastily-formed and ill-founded conclusions, Mr Brancker then got down to the serious business of thrashing the Government for its deceit in "completely altering the conditions" under which the Public Inquiry had considered Stansted as a third airport site.

He recalled that the Inquiry had heard evidence about a proposed airport with one pair of parallel runways, each 12,000 feet long, with a third, much shorter runway.

But here was the White Paper talking in terms of an airport with three pairs of runways to cope with the forecast traffic demand in the 1980's. Such an airport, the White Paper blandly explained, would need to be four miles wide. Moreover, to obtain a simple system of taxi-ing and ground movement generally, it might well need to be seven miles long!

"This possibility was never examined in detail at the Inquiry. The White Paper states quite categorically that Stansted can be expanded to provide two pairs of runways, but no evidence whatever to this effect was produced at the Inquiry."

Mr. Brancker quite clearly felt the Public Inquiry a complete and utter waste of valuable time and money when a White Paper could so blatantly disregard and misinterpret its findings—then attempt to get away with it.

In his report to the Government, Mr Blake, the Inspector at the Public Inquiry, had recommended a further study of the whole airport problem. The White Paper claimed this had been done. But it added that an independent commission had been considered "unnecessary" because of the Inter-Departmental Committee's "thorough examination" in 1962 and 1963. The subsequent publication of the Committee's report—and the Public Inquiry itself—had given "an ample and well-used opportunity to all interests" to put their point of view.

If ever there was a bid to get an airport built via the "back-door" method, this was it. Mr. Brancker seized on it immediately and stormed back: "One thing which the Public Inquiry did show was that the Inter-Departmental Committee had not made a thorough examination of the question.

"With the exception of the viability of Stansted in terms of air traffic control, and the fact that it was physically possible to build a pair of parallel runways on the site, almost every other bit of evidence came to pieces under cross-examination.

"This statement, therefore, is wholly erroneous, and cannot be accepted as a reason for not having an independent commission. On the point of an independent commission being time-consuming, it must be pointed out that the re-examination by the Government itself took one year and the national planning authorities have still not been consulted."

The "re-examination" of the proposal was ordered by Mr Douglas Jay, then President of the Board of Trade. To this day, nobody knows what form the review took, who sat on the panel, or how many alternative sites they considered. Many people still firmly believe that among its members were those who formed the Inter-Departmental Committee.

If this was so, it could hardly be considered an independent judgment. The Stansted Preservationists tagged it "The Star Chamber Review" because of its secrecy—and nobody has contradicted them so far!

Mr Brancker agreed with the White Paper decision that a number of the examined sites had been abandoned for one reason or another. For instance, there was not much point having an airport to serve London located on the south coast.

He further accepted the view that Stansted came out top of the class from the air traffic control aspect. But the position of the Thames Estuary sites was not so straightforward. He recalled that, at the Inquiry, much evidence had been heard about possible sites at Sheppey and Cliffe—both on the east coast. But nothing had been said about Foulness—considered "an impossibility" because of the military firing range at Shoeburyness.

Furthermore, the White Paper said any airport at Foulness would mean the closure of the locally owned Southend airport. According to Mr Brancker, that might not be a bad idea. After all, he explained, although it was a busy airport and made a profit, there were already problems about the noise from jet aircraft:

". . . in the long term, Southend might be better off if the present airport area was devoted to other purposes and the business transferred to a fully developed airport at Foulness, only a few miles away. There is no evidence of this solution having been examined in any detail—presumably because of being completely inhibited by Shoeburyness."

Although he did not say so in so many words, Mr Brancker must have had Foulness partly in mind when he pointed out: "The third airport is going to be a national asset or liability for a very long time and will still be used for civil purposes when defence requirements may have changed radically

"Consequently, there is a risk in permitting immediate military needs to over-ride the long-term transport necessities."

He then decided to attack the White Paper and the Inter-Departmental Committee for not being able to calculate time and distances very accurately. The Committee had rejected some possible sites because they were more than an hour's drive from Grosvenor Square. But the Public Inquiry showed conclusively that Stansted was in the same situation. Therefore, it enjoyed no special advantage in this respect. Consequently, there appeared to be a good case for re-examining some of the earlier rejects where access was considered a major objection.

Then there was the problem of aircraft noise. The White Paper said noise nuisance would not be as bad at Stansted as the Inquiry report suggested. There was not a shred of evidence for this statement.

Further, the Paper's claim that the number of people at Stansted who would be subjected to very high noise levels was only five per cent of those affected at Heathrow, had not been tested by cross-examination at the Inquiry.

The access problem to Stansted had been widely examined at the Inquiry. It had been conclusively proved it was impossible to reach the airport site in under an hour from Grosvenor Square —although this was one of the reasons the Inter-Departmental Committee had selected it in the first place!

Evidence had produced various time estimates of between 65 and 150 minutes—and even these depended on major road improvements in addition to the completion of the M11 motorway. Yet the White Paper, without any kind of evidence, stated that the off-peak journey time would be seventy minutes. Where did it get its figures from?

This was a further example of where the White Paper had set out entirely different conditions to those examined at the Inquiry. Then, all the argument and graphs of noise disturbance had been based on a single pair of runways. Two pairs would radically change the picture. Areas not considered at all would now be subjected to high noise nuisance.

Mr Brancker was not finished with the White Paper yet. He attacked it further on costs. It had claimed that to build an airport at Stansted, with two parallel runways, was £47 million. Additional to this would be the cost of replacing facilities at the USAF base at nearby Wethersfield. But the estimated figure did include the provision of access by road and rail.

Where did the figures come from? No cost figures had been presented to the Inquiry and, in any case, it was not clear whether the £47 million included the cost of obtaining the land required immediately—for the two-runway airport—or for the future four-runway terminal.

The impact of an airport at Stansted would be "very much" greater than had been anticipated, Mr Brancker warned. There was not only the question of considerable population growth caused by people moving into the area for jobs at the new terminal —but the inevitable industrial growth. Despite this, the national planning authorities had been given no opportunity of stating their case at the Inquiry because they were not directly represented.

Mr Brancker summarised his "analysis" with this unprecedented condemnation of an official Government document:

"One of the more disconcerting aspects of the White Paper is that it has doubled the potential size of the project without giving very much weight to the possible effects of such things as noise, access, industrial development, land required, etc.

"It may be wise to stave off the need for a fourth airport by making the third expandable, but this must surely imply even greater care in selecting the right site. In addition, the Government seems to have retreated from the minimum conditions which it imposed in the first instance and on which the choice of Stansted was predicated.

"As a public document, the White Paper fails to give the impression of being impartial. This may be entirely due to the manner in which it is written. But considering the size and importance of the project and the expensive difficulties which have arisen in the past in this context, a more dispassionate costing of alternatives would have been reassuring.

"The third London airport cannot be considered in isolation. So its location must inevitably form part of a broad development plan. Some very unorthodox things may well have to be done to find an imaginative solution which will be of value in the future.

"Stansted may ultimately be found to fit into such a plan, and to be the only place that does. But, until the alternative solutions have been examined in depth—and the comparative costs are available—I cannot feel that the case is proven."

Rarely, if ever, has a Government case been so thoroughly shot to pieces—particularly by its own acknowledged technical expert. It was certainly a slap in the eye for the Whitehall bureaucrats. How could any official case survive after such a thoroughly logical but merciless pounding?

CHAPTER SIX

National Airports Plan

One thing the third airport controversy highlighted was Britain's serious lack of a national airports plan. Where, for instance, should our air terminals be located? Should there be just a few, selected large bases, or a string of smaller airfields dotted around the country? Which would suit the national requirement better? And who should own them?

Mr Brancker's views, in no way critical of what any Government had done so far, were highly relevant in relation to the future growth of the air business in Britain.

He thought, for instance, that the handing over by the Government of certain major airports to the British Airports Authority, while "wholly logical", gave the unintentional impression that others outside its control were relatively unimportant. The move had the further hint that—apart from more runways to serve London—no others were necessary.

But Mr Brancker questioned the truth and wisdom of this policy. Although no answer to the problem was ready-made, there was no doubt that, in view of the growing air travel industry, more airports would be needed in the future.

The growth of general and executive flying in Britain had been slow compared with the United States. But it was not impossible that this had been due to the lack of suitably located airports.

"Patently, it is impossible to wave a wand and produce airports immediately, but it is important to know where they should be", he explained. The Stansted problem, for example, would have been much easier solved if viewed in the context of a national airport development plan.

"It must, at all times, be remembered, that air transport

has now ceased to have luxury connotations; it is as basic a method of transport as rail and road. This is not to say it serves the same purposes—but it serves the same people.

"Failure to develop it fully—within sensible economic limits —will be just as harmful to the national interest as to neglect the road system or to limit the use of British shipping."

Although London was, at the moment, the centre of most air traffic in Britain, it was misleading to assume the same situation would always exist. Centres of population and industry might change and, unless traffic was forced into artificial channels, these areas would need air services.

Still, at the moment, London was the air funnel for Britain. Even people going abroad from provincial centres were, at the moment, regularly channelled through the capital. This was partly due to traditional habits. It was also partly due to the fact that London had many more international routes than, say, Manchester, Glasgow or Birmingham.

"In some respects, we have started a vicious circle: London is popular because it has frequent services; and, because it is popular, more frequent services are provided."

According to Mr Brancker, the circle was partly the fault of the airlines. They were looking for quick profitability and London provided the answer. "Consequently direct services from the provinces to points abroad have tended to be neglected, primarily because they did not offer much prospect of quick profitability."

This was important in the context of a national airports plan. Britain could not easily support a large chain of airports unless they were used sufficiently to make them economically worthwhile. Therefore, one of the first tasks implicit in any grand design would be to discover just how much business was available.

For instance, a different approach to the licensing of routes —with a clear intent to see the provinces were properly served— might change the picture considerably. Under such a scheme, airlines would not be allowed to operate the more profitable routes unless they accepted some of the less attractive ones.

But any national airports policy, according to Mr Brancker, must be seen in relation to a general concept of British aviation as a whole. Otherwise, it would be like planning a railway without knowing where the stations were to be, or without locating the junctions or marshalling yards. Any approach must be realistic.

By any standards, building major international airports was an expensive item on any budget. They could not, therefore, be constructed in large numbers. "The ultimate number and pattern will not only depend on the size of the traffic demand, but on the area which can be conveniently tapped by surface transport and feeder services. Any plan must recognise the possibility of new and unorthodox means of surface transport."

The future demand, Mr Brancker considered, would be for airports capable of handling inter-Europe traffic. On this basis, he thought it might be worth starting from the base that any conurbation of more than one million people should not be more than thirty minutes driving time away from such a terminal.

Clearly, this could not be done in every instance. But the object must be to ensure that the nation, as a whole, was adequately served. Further, places which had inter-Continental airports would not require separate European terminals.

The first essential was need. Coupled with this should be future changes in population and industrial centres. Studies along these lines would show which centres were well-served at the moment and where airports would be needed in the future.

"Wherever possible, each new airport should form part of an industrial complex, so that the problem of noise in residential areas is avoided. Siting must fit as closely as possible into local development schemes," Mr Brancker continued.

He stressed the importance of choosing runway sites which "did not disturb the community". This, mainly, was because of the noise factor which, apart from any social considerations, would result in future development being restricted.

Even when such a plan had been formulated and suitable sites chosen, there remained another problem—ownership and administration. At present the system was rather slapdash. Some airports were controlled by the BAA, others by local authorities and some directly by the Government.

"It might appear rational for all airports used by regular services to be brought under one authority, but the most productive policy might be for all new airports to be taken over and used by the BAA, leaving existing arrangements unchanged."

But Mr Brancker warned that it would be wishful thinking that any new airport would become economically viable overnight. In fact, for the first few years they might need subsidies, both from the Government—and the local communities which would eventually benefit most.

It was worth remembering that the "tremendous" development of domestic and international air services in the United

States had happened only after considerable financial assistance had been given by the Civil Aeronautics Board. Local services were still assisted in this way, although the trunk services were totally self-sustaining.

The same thing should happen in Britain—mainly so that landing charges and other fees could be kept at a reasonable level. This should continue until sufficient numbers of people used the airport to make it financially sound.

Why has Britain lagged so far behind in the development of airports? One reason, according to Mr Brancker, could be the belief that Vertical Take-Off and Landing, and Short Take-Off and Landing planes would be developed in the near future. These, of course, would require less space than conventional terminals.

But he has an answer to that. It would happen in due course, but large numbers of orthodox aircraft were still being developed, produced and delivered. These would be in service for at least the next twenty years and needed conventional runways. Naturally, their needs could not be disregarded.

Even when Vertical Take-Off planes had been developed to the point where they took over from existing aircraft, the extra available runway area need not be wasted. Indeed, "if airports are developed for what they really are—transport centres—then these spaces will have considerable value when the time comes to release them".

There was also a case to be made for small landing strips in places which did not need a full-scale airport. These would be suitable for private or executive flying. These planes would be allowed to use main airports—provided they carried the right equipment.

"It is important, however, for a start to be made as quickly as possible to set out a plan for airports on a national scale. Apart from basic research—which is imperative—consultation should take place with all interested parties, particularly the airlines.

"Unfortunately, the Air Transport Licensing Board does not possess the basic store of traffic information which has been built up over the years by the Civil Aeronautics Board."

Although some dislocation, in any local area, was inevitable, it would be kept to a minimum if a careful study was made of every alternative site before a decision was made and construction started. Existing airports should be integrated as closely as possible with any future plan, providing they met the required standards.

Mr Brancker emphasised that, providing access was good, the numbers of airports should be kept as small as possible. This would tend to give more frequent services to more destinations. Increased use would also improve economic results.

The present process under which air licences were issued to operators needed to be reviewed. Although provincial routes in Britain had not been sufficiently developed, this was not the fault of the Licensing Board, as it was presently constituted and directed.

But—given the right directive and staff—it could "work wonders" assuming it was allowed the right machinery to determine basic demands. This would be a necessary requirement to any provincial airport development scheme.

"The problem is urgent. The sooner a study is started and the basic needs determined, the easier it will be to mould airport development with general industrial programming and growth. The later it is left, the more difficult it will be to find sensible integrated solutions and the country will be faced with a series of 'Stansted' crises."

Although it would obviously not be possible to build all the required airports immediately, explained Mr Brancker, it was important to start now. The primary task was to determine exactly where they should be, then plan round them. The advantage of this would be that road and rail development could be geared to take account of the new needs.

Any such plan should be implemented by the BAA—particularly as some of the new airports would need financial assistance.

"One thing must be stressed: the object of producing a national airports plan is not to build airports for the sake of building airports.

"It is essentially a transport plan to decide as soon as possible what future requirements will be—and how best they can be met.

"If the ingenious use of surface transport can reduce the number of airports needed, so much the better. But it is essential that the country is prepared for the future."

CHAPTER SEVEN

Roskill: High Drama or Farce

In the end, something had to give. And it was Parliament. On 20 May 1968 Mr Anthony Crosland, the President of the Board of Trade, announced that there would be a large-scale Inquiry into the siting of the Third London airport.

He said that there had been discussions with the Tory opposition and he acknowledged the constructive help that had been given from the other side of the House. "We have reached a broad agreement", he said.

Mr Crosland said that the Inquiry would be conducted by a non-statutory commission headed by a High Court Judge. And he gave as its terms of reference:

1. To inquire into the timing of the need for a four-runway airport to cater for the growth of traffic at existing airports serving the London area, to consider the various alternative sites, and to recommend which site should be selected.
2. General planning issues including population and employment growth, noise, amenity, and effect on agriculture and existing property; aviation issues, including air traffic control and safety; surface access; defence issues; and cost, including the need for cost-benefit analysis.

He named the man to head what was termed "The Inquiry to end all Inquiries"—Mr Justice Roskill, an eminent High Court Judge who was to have a six-man team.

Mr Crosland underlined Mr Justice Roskill's task. He told the House: "The form of the Inquiry must meet two requirements. On the one hand, this is one of the most important investment and planning decisions which the nation must make in the next decade. This points to an expert, rigorous and systematic study of the many and complex problems involved.

"At the same time, the decision will profoundly affect the lives of thousands of people living near the chosen site. This calls for an adequate method of representation of the local interests affected."

The move was welcomed from the other side of the House. Mr Frederick Corfield, then Opposition spokesman on aviation affairs and now the Tory Government's Aviation Supply Minister, said that he welcomed the approach being adopted by the Government.

The Opposition also welcomed the appointment of Mr Justice Roskill as Chairman along with "the apparent intention that there should be a senior planning inspector of the Ministry of Housing as a member of the commission". Mr Corfield quizzed Mr Crosland about the membership of Roskill's team. What sort of skills or expertise would be represented on the commission? He also asked if the commission would be free to take evidence from the Ministry of Defence. This was obviously a long shot, from Mr Corfield's point of view, that a coastal site might be chosen or short-listed along the Thames estuary, with the implications that it could be affected by the top-secret Shoeburyness firing range.

On his part Mr Crosland gave an exact and precise answer —one that satisfied members on both sides of the House. He said: "We think the kind of membership in mind would include perhaps a traffic engineer, aviation expert, economist, a businessman, and a regional planner, in addition to the planning inspector".

The Ministry of Defence would be submitting evidence almost certainly. But it would be of a classified character and this would have to be considered *in camera*. Then Mr Crosland again underlined the importance of the Roskill Commission's work. He said bluntly that it would take at least two years to report.

Both the experts and the critics were quick to seize onto the amount of time that Mr Justice Roskill was to be allowed to sift evidence and prepare his final report. It almost certainly meant that the site eventually chosen by the Inquiry would not be ready by the time both Heathrow and Gatwick airports reached saturation point. And according to the estimate of the British Airports Authority that would be in 1974. It is noticeable that this date was stretched by the BAA later in evidence to 75/76.

Sir Ian Orr-Ewing, Conservative MP for North Hendon,

asked Mr Crosland whether the timing for a third London airport depended on the way in which the traffic could be handled at Heathrow and at Gatwick. And was it within the terms of reference of the Commission to probe methods of improving air traffic handling capacity at the two airports? This should be pushed hard before the country was asked to invest in a new centre of noise and nuisance.

Mr Crosland politely corrected the query by saying that the terms of reference of the commission included "the timing of the need" for the third, four-runway airport. "So I think this point will be covered."

Mr Crosland did his best to take the word farce out of the long and miserable search for London's third airport site. He told the House that the reason for the delay in announcing the setting up of the Commission was because talks had been held with the Opposition to get their point of view and added: "In view of the tangled history of this question, it is extraordinarily important to find the right kind of Inquiry, and partly because I have consistently taken the view that some delay is to be expected and is well worth incurring if, as a result, we get the right decision and one which everyone admits to be right."

The ball then appeared to be wholly in Mr Justice Roskill's court. At the time of the announcement he sat in the Queen's Bench Division in the High Court and was Vice-Chairman of the Parole Board for England and Wales. He was a judge in the dispute, a little beforehand, involving a £1,200,000 ship deal between Mr Panaghis Vergottis, a Greek shipowner and Maria Callas, the prima donna, and Mr Aristotle Onassis.

Roskill was to examine 48 sites and would be helped by the British Airports Authority, which had already surveyed many within 100 miles of London. But it was made clear that the Authority would not attempt to lead the Commission but merely to try and save time, although the probe would last two years anyway.

This was 32 sites less than the number claimed to have been scrutinised by the secret review body, whose findings form the basis of the White Paper of 12 May 1967, which recommended that, after all, the airport should be built at Stansted. The Government's decision to appoint the Roskill Commission was unanimously applauded by the Press.

Said the *Sun* newspaper: "No half measures about Mr

Anthony Crosland. Having accepted the need for a fresh Inquiry into the siting of London's proposed third airport, he is now going the whole hog."

And then hugging itself with selfrighteousness the *Sun* added: "Right up until February the Government were insisting —despite much contrary evidence—that Stansted in Essex was the best site, the cheapest site and indeed the only possible site unless there was to be a fatal delay.

"Too late to change now was the cry with which reluctant Government MPs were whipped into support. Now Mr Crosland, who inherited the problem from Mr Douglas Jay, sees it very differently and the *Sun* is delighted that he does."

The Times, a little more soberly, said: "The procedure is untried and the wide range of consideration which Mr Crosland rightly places before the commission, including the need for cost benefit analysis, requires much from its members.

"But the opportunity is now provided of openly examining the case Civil Servants have been making, repairing its suspected deficiencies, making a true comparison of costs, giving due weight to factors other than aviation and allowing those affected a fair hearing—in short, of removing the objections of the Stansted objectors."

The Times observed that if there was some delay it was a reasonable price to pay for getting the airport in the right position.

The Daily Telegraph said tartly: "What has become, now, of the rigid demands of the British Airports Authority for a start on Stansted at the earliest possible moment, in order that London could cope with the swelling air traffic expected by 1974? Clearly, the Authority will have to wait; the stubbornness with which the former President of the Board of Trade Mr Jay defended the Stansted scheme on the grounds of urgency, has been blandly brushed aside. That which was immutable and, so they said, technically inevitable, is mercifully now open to independent valuation. All this is to the good. The claims for a really large airport in the Thames estuary can now be thoroughly examined and those claims are very strong. The Stansted Preservation Society can feel justifiably proud of itself. So there is now a good chance that the right rather than the easiest solution will be found to this problem."

The Government, and Mr Crosland were delighted with the reaction. It could be argued that the terms of reference given to

Roskill were so wide—even so generous—that it left very little come back to the Government of charges that it had over-ridden public interest in siting the airport.

The Roskill Commission got down to the hard work a month after the announcement. But it took ten months, to 4 March 1969, before it came up with its first main prize—a short list of four sites. They were Foulness, on the Essex Coast, Nuthampstead in Hertfordshire, Cublington in Buckinghamshire, and Thurleigh in Bedfordshire.

Perhaps the biggest surprise was that Roskill had dropped Stansted in his prior investigations. Perhaps the greater part of the surprise had come three weeks before when, in fact on Saturday morning of 15 February, two national newspapers obviously found out the important part of the Roskill Commission's job.

Both *The Times* report by their air correspondent Mr Arthur Reed, and the *Daily Express* report by Mr Frank Robson, had made it clear, or virtually made it clear, that Stansted was not to be among the four short-listed sites. To the Preservationists it was almost unbelievable that Stansted should be missing. Their long and tiring battle had at last been won. They had stopped a Government dead in its tracks, forced a change of mind, set up a vital commission of Inquiry and then finally erased Stansted from the running.

The nineteen square miles of agricultural land which, for 400 years, had been producing some of Britain's finest vegetables and wheat crops, would now, thankfully, remain undisturbed by the bulldozers and airport-making equipment. There were champagne parties in Stansted and the surrounding little villages on the night of 4 March when the short list was announced. It was seen not as a reprieve any longer but a complete victory. And suitably the celebration had to be underlined in champagne.

What was a surprise however was the distance from London of the four chosen sites. Cublington, sometimes known as Wing, is about 41 miles from London. Thurleigh in Bedfordshire is 54 miles from London. Nuthampstead in Hertfordshire, and the fourth on the official list, Foulness, 52 miles.

The sites:

Cublington is close to Leighton Buzzard and the electric railway to Birmingham and Lancashire, and is 12 miles from the heavily-used M1 motorway.

Thurleigh lies to the north of Bedford. The site is close to the under-used main line railway from St Pancras to Leicester, Nottingham and Yorkshire, and seven miles to the West of the Great North road a dual carriageway but with room for improvement.

Nuthampstead to the north-west of Stansted and on the Cambridgeshire-Hertford border had the advantage of the proposed M11 motorway which would run to the east of it, as does the existing main line to Cambridge.

Foulness on the Thames estuary, off the Essex coast, was immediately thought of as a site built on reclaimed land. It could be served by an extension of the railway line but would call for a further rail link and major road works.

Mr Justice Roskill said in the letter to Mr Crosland, President of the Board of Trade, exactly why Stansted had been missed off the list. "You'll notice that Stansted does not find a place, while Nuthampstead has been included. In the Commission's view of the possible sites in that area, Nuthampstead offers most advantage over Stansted, particularly in respect of noise and of air traffic compatibility with Heathrow."

So noise and air traffic factors, almost certainly ruled out Stansted. Without doubt another deciding factor was the feeling that the Stansted question had become much too emotional. There was never any Government pressure put on Roskill to leave Stansted off his short list. He was given complete independence under his terms of reference and, according to the Commission's members and subsequent reports, it was left off after a minute examination of its advantages and disadvantages as an airport for the late seventies, handling big jets and possibly supersonic airliners.

However Nuthampstead was a mere six miles from Stansted. On 4 March, the day the report came out, Mr Douglas Jay was philosophical about the rejection of Stansted, a project about which he had fought so hard and which had partly led to his leaving the Board of Trade.

They haven't scrapped it, he said, they are merely calling it Nuthampstead. And as the people of Stansted rejoiced, because rejoice they certainly did, campaigns began almost immediately among the residents in the areas of the four short-listed sites. From the day of the announcement, Mr John Raven, the chairman of the Wing Rural Council, which covers Cublington, pledged: "We are going to fight this all the way." He called the suggestion that Nuthampstead or Wing could become, or might become, London's third airport, "a tremendous bomb-shell."

And while areas newly short-listed were upset at the prospects of having an airport of the future on their back doors, the British Airports Authority was upset for a completely different reason.

The BAA, like Mr Douglas Jay, had been, in fact, quite desperate to have Number Three at Stansted. Each of the four new sites was significantly more distance from London than Stansted, the Airports Authority stressed. It also expressed incredible surprise that Stansted had been dropped from the list. Even now the Authority's experts read the first part of the Roskill report in utter disbelief.

Clearly, what the British Airports Authority could not understand was how the Roskill Commission could publish a short list without even including Stansted. The feeling of the Authority and its Chairman, Mr Peter Masefield, merely reflected a thousand times over the fantastic victory by the people of Stansted and by the campaigners against the airport. But after taking a little time to recover from the shock, the Authority quickly made it clear that they were not going to take the decision about Stansted completely lying down. There was an airport there, the airport built by the American Air Force during the war. The airport belonged to them. It was a British Airports Authority preserve, as was London's principle airport Heathrow, London's Gatwick and Prestwick in Scotland. So they felt fully entitled to develop the airport within its existing boundaries, and said that they would do this, although Stansted would not be developed as a major international terminal for scheduled traffic. The aim was to push more and more charter work into Stansted, not just cross-channel flights or short inclusive cut-price tour holidays to European travel sun spots, but also transatlantic flights with big jets.

Immediately the opponents to the airport being sited at Wing, Thurleigh and Nuthampstead joined forces and their plea was that Foulness was the only acceptable site. Mr W. J. Bandy, chairman of the parish council of Wing, declared: "We are horrified at the prospect of the third London airport coming here. During the war there was a bomber command aerodrome near here and they were noisy nights then. Jet aircraft overhead day and night would be absolutely intolerable."

Mr Bandy said the Wing Rural District and the Parish Council's Association was seeking a consortium of the three inland sites to press for Foulness as the only acceptable place, on the ground of the least population disturbance, and the fact that much of the noise would be shed over the North Sea by the big jets.

Mr Robert Maxwell, then MP for Buckingham, said that the opposition would make the Stansted fight look "like child's play ... I'm appalled at this proposal because north Buckinghamshire, one of the loveliest parts of the home counties, has already been made to give 22,000 acres for Milton Keynes, to help rehouse Londoners. We have also been asked to give up several thousand more acres for reservoirs.

"It is unreasonable and unfair that Buckinghamshire should be made to give up extra land, and what about all the inconvenience to the people who live there?"

A Bedfordshire airport resistance association was formed to oppose the choice of Thurleigh. There was already a 10,500 foot runway, 300 feet wide, which was in use by the top-secret Royal Aircraft Establishment at Bedford for aviation research. Many millions of pounds had been put into this research establishment by the Government. Officials at the establishment would say nothing about the Roskill proposals but it was clear that much of the vital work being done there, particularly in wind tunnels, probably the best in Europe, would have to be transferred elsewhere. This would involve enormous cost and then again where would *this* establishment be resited? It could not be linked up with its sister establishment at Farnborough. It was much too big.

In the pretty and tiny hamlet of Nuthampstead, with a population of 108, just to the south of Royston, the locals had been preparing for battle since the premature reports that Nuthampstead would be on the Roskill Commission's short list.

Again, during the war, there was an airfield used by the US Airforce, but it reverted back to agricultural use in 1946.

In the brief time since the news had "leaked" in February, a co-ordinating committee had been formed and finance promised to fight the airports proposal. At the time, Mr Donald Robertson, chairman, said that every medium and argument would be used to show Nuthampstead to be even less suitable than Stansted. He made a special case out of its proximity to Stansted, perhaps supporting Mr Douglas Jay's view that Nuthampstead was really just Stansted in another guise.

At Luton, home of another airport, Mr Derek Samson, president of the Luton and District Association for the Control of Aircraft Noise, and a Fellow of the Royal Aeronautical Society, said: "I feel it is fantastic that Nuthampstead should be considered. The only rational place is on the coast at Foulness, and this is the view of my Association." Residents at the three inland sites all simultaneously pointed their finger at Foulness, highlighting its great advantage as an airport site. Oddly enough Foulness was to be the first site at which a public hearing was to take place, starting on 5 May 1969. Hearings for Nuthampstead were set for 9 June, Cublington on 14 July, and Thurleigh on 8 September.

And whilst these "stage two" local hearings were in progress the Roskill Commission was getting on with its stage three task —undertaking the detailed research into the short-listed sites. It was hoped that that would be completed by the end of 1969 and the Commission wanted all those who wished to give relevant evidence to do so by 31 July of the same year. The County Councils in each of the areas short-listed, that is Hertfordshire, Bedfordshire, Buckinghamshire and Essex, were each to campaign to have the airport built at Foulness. The odd exception here of course was the County Council of Essex, who were backing the moves to have the airport built within its own boundaries.

The arguments flowed just as they had at the time of Stansted: that, by building an inland site, valuable farm land would be used and transformed into a concrete jungle. Just one of the many phrases used at the time: "A concrete jungle of runways and aprons and buildings."

Apart from that there would be the danger of aircraft circling over busy areas, the nuisance of aircraft noise, whereas it was pointed out repeatedly (as it was to be pointed out many more times at the lengthy discussions held by Roskill) that no land would be required at Foulness to build the airport. The area in question was to be created, to be reclaimed from the sea. This would mean also that aircraft would spend much of their time while waiting to land circling over the sea, or partly over the sea, and that they would also in many cases take off over the sea. From the point of view of eliminating danger, this must surely be in favour of a coastal site.

Although the choice of Nuthampstead had come as no real surprise to the locals, the people in the area were only just getting used to the idea of it possibly becoming an airport again. When American Flying Fortresses ceased to rumble down the concrete ribbons of Nuthampstead, local farmers cast envious eyes at the vast spread of land lying fallow. Eventually, $13\frac{1}{2}$ years ago, the Government freed the land and the back-breaking struggle began of trying to reclaim it, and one of the men who did a great deal to put it back into productivity was Tony Barker.

He realised immediately that if the third airport was to be built in that sleepy backwater bordering three counties, he would be the biggest loser, for the whole of his highly mechanised 400 acre Bulls Farm, would vanish.

"It's been damned hard work, ten years or more of it," he said. It had cost him £15,000 to rip up all the old runways and get the land back under the plough. Similar complaints to that of Tony Barker came from other parts of Nuthampstead and Thurleigh and Cublington. They echoed the complaints that had flooded in from the people in the Stanstead area.

At Foulness it was different. Most of the land bordering the mudflats, which might be reclaimed to form an airport, was marshland. Clergymen at Stansted had been high on the list of protesters about what aircraft noise might do to old ecclesiastical buildings, the beautiful churches. Now it was the turn of the vicar of Wing, the Rev Geoffrey Willis. Both churches there, he said would be shaken to pieces by noise and vibration from low-flying aircraft.

His church at Wing, he said, had survived a Danish invasion and other battles during the 1,300 years since it was first built, but an airport there would certainly mean the end.

Although there were few objectors at Foulness, one of the most vocal had been the Ministry of Defence, whose much-mentioned and highly secret Shoeburyness artillery range could not, they said, be moved. But now an alternative site was found in South Wales. This removed the big obstacle to the building of an airport on the miles of Maplin Sands, and undoubtedly influenced Mr Justice Roskill's decision to include it in the short list. The result of the move was that protests over the siting of London's third airport spread to the South Wales coast, because although the Welsh were in no way disturbed by the airport, they did not want the quiet of their countryside blasted by the continual firing and testing of 25 pound artillery shells.

The final phase of the Roskill Commission began in a downstairs banqueting room at the Piccadilly Hotel, in London's West End, on 6 April 1970. It was to last for 74 days, and when it closed on 12 August, more than 160 witnesses had given evidence amounting to more than three million words. It was the longest Public Inquiry of its type in history. It was also, one of the most important ever to be heard, particularly in view of what had gone before. For not only had the High Court Judge and his team of experts got to ensure that justice was done, but that it was seen to be done. The public were, however, in no mood to listen to the often complicated expert evidence and the long arguments, largely technical.

Space had been allocated in the room for more than fifty members of the Press who, in the past, had helped stir up the row over Stansted airport. But most days, four or five journalists at the most were present. There was also plenty of room for the public. But most of their benches were empty. Only one man, Mr W. J. Lewis of Bridgwater, Somerset, attended every session. He said afterwards that he was fascinated by the Inquiry because his special interest was civil aviation. He became such a well-known figure, that even Mr Justice Roskill, in his summing up, commented on Mr Lewis's unswerving interest.

When, on 12 August, a tired Mr Justice Roskill completed his long and highly technical task, he leaned back, lit a cigarette and said: "Now I understand they will be having this room back for bingo and bridge." It was then that the forty parties of barristers, solicitors and expert witnesses, who throughout the Inquiry had been provided with special private dining facilities and other amenities, returned to resume their normal practices.

But, although many of the legal experts and technical witnesses had been at loggerheads during the hearing of evidence, it was not all hostile. On 14 June, for instance, the Commission's own research team turned out for a cricket match against the objectors from Foulness and, on another occasion, Mr Douglas Frank, QC, for the Essex, Hertfordshire and Cambridgeshire County Councils, led his own cricket eleven.

Many important libraries and universities had already placed firm orders for the Commission's three million words of evidence. For, not only were they an intrinsically historic document, they had formed the basis for collating much hitherto unknown information which would be vital for any future large-scale planning development. For instance, almost all of the first month of the Inquiry was taken up with a close examination of what had been generally referred to as a cost/benefit analysis. After the first phase of the Inquiry, the Commission in announcing its four sites, decided that although Foulness was an important possibility, it was on this cost/benefit basis that it was found to be most lacking. But, after many days of cross-examination of technical experts, it became more than clear that any yardstick by which cost/benefit of any area could be analysed, was bound to be incorrect. It was apparent that the whole foundation of any such analysis was unscientific and difficult to argue.

How, for instance, do you equate the value of a Norman or Saxon church with its replacement by an airport runway or terminal building? It is now self-evident that you cannot measure what you are going to lose in architectural beauty and historic value. *For all its immaculate and dedicated investigation, in the years to come, many will ask the question: Was Roskill really necessary?*

Foulness, to many of the witnesses who had appeared before the Commission, seemed to have a half chance of selection. But even here there were strong arguments against. No matter which site Roskill chose there was bound to be wide opposition. The saving grace about Foulness was that only a handful of people would be affected by the construction of an airport—and even a deep-water sea port—on the sands at Maplin. Even in mid-stream Roskill came across another problem. Along came two major companies vying for a place in the sun, with a massive industrial port and airport complex at Foulness.

These industrial giants claimed that, with big city backing they could save the taxpayer from footing the bill for an airport. Although estimates have varied widely, the latest indicate that a four-runway airport would cost in the region of £110 to £150 million—a substantial saving to the taxpayer in view of the new Tory Government's pledges to cut Government spending, and thereby reduce taxation.

A free airport is an attractive proposition to anyone, and many people now feel that Foulness should be built—just because it would be free. The "let us build a sea port and you can have an airport for nothing" consortiums made the bait sound tasty. Whether such an airport would, in the long run, be entirely gratis is a matter of some speculation which will be discussed later in the book. Mr Peter Masefield, chairman of the British Airports Authority, believes the financial costs will still be a heavy burden to the country.

Mr Justice Roskill said that he hoped his Commission would complete its report to the Government by the end of the year. But later it became clear the completed report had slipped into early 1971. From the Government side Mr Frederick Corfield, Aviation and Supply Minister, spelt out the Government's intentions when he visited Gatwick in October of last year.

In mentioning the Roskill report he said that the Government would note the Roskill recommendation but that did not hold the Government to accepting the recommendation. The Government still had the right to select *one* of the four sites examined by the Roskill Commission. So in the end the Government was to have the last word. Faced with this, it is not surprising that, on the day the Inquiry closed, one leading counsel said that after all the Commission's hard work the final chosen site for London's Third Airport could still go down in history as "Roskill's Folly".

CHAPTER EIGHT

The Airlines: The Almost Silent Voice

Throughout the protracted and rowdy battles surrounding the selection of an acceptable site for London's third international airport, one important voice was strangely muted: the world's airline operators.

Many, especially the two State-owned British companies, had been expected to express firm and strong views on which site they preferred. After all, it was realised, a united stand by them would have been a powerful factor in swaying public opinion since they would have to use the new terminal more than anyone else. But they kept a discreet official silence.

Some of the large foreign airlines had been patiently waiting for BOAC and BEA to take a lead in opting for one place or another. It would then almost certainly have become a game of follow-your-leader, with the overseas carriers falling in behind. But the British preferred to keep quiet until a firm Government decision had been reached.

Furthermore, they claimed they had done no political lobbying behind the scenes. Their only contribution, in fact, had been to give evidence to the Roskill Commission, mainly non-committal, on the difficulties of running a major airport.

Their attitude was not really surprising. Both the British operators and some foreign airlines had major capital investments at Heathrow. BOAC, alone, had spent nearly £40 million by the spring of 1971—much of the recent expenditure being for engineering workshops and hangars for Jumbo jets.

No company with that sort of money invested in one place would willingly pick up its bags and move elsewhere. BEA was, basically, in the same position.

At the same time, they, and many other airlines, were dis-

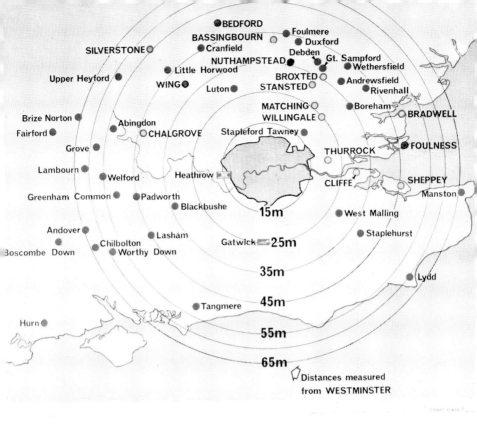

Here is the real extent of the threat: the chart indicates forty-eight alternative sites for the third London Airport, each of which was investigated thoroughly.

[*By courtesy of the British Airports Authority*]

This is Stansted Airport in the heart of some of the most beautiful countryside in England. This is the area—and more—which would have been expanded into London's largest air terminal.

Right. *Two typical aspects of the grass roots campaign against the siting of the third London Airport in very beautiful and agriculturally valuable areas.*

This £300,000 scale model of the Thames estuary was constructed by Tedco, one of the groups interested in developing the Foulness area. Its main objective

was to show the effect of tidal movement, particularly to ensure that any airport or seaport would not cause flooding dangers to London.

Right. *This is Foulness and the Maplin Sands. This is where the third airport is to be built. The tracts in the centre of the picture are made by the military vehicles which recover artillery shells, fired into the sands from the nearby Shoeburyness testing ranges.*

[*By kind permission of the Daily Telegraph*]

Left. *The Roskill Commission, made up from left to right by Professor A. A. Walters, Professor David Keith Lucas, Professor of Aircraft design, Mr A. J. Hunt, principal Planning Inspector, Ministry of Housing and Local Government, Mr J. Cairnes, Secretary, The Hon. Mr Justice Roskill, Chairman, Mr A. Goldstein, partner in R. Travers Morgan and Partners, Consulting Engineers, Professor Colin Buchanan.*

Mr Bernard L. Clark (left), *the man whose dreams will come true when the airport is built at Foulness, explains his scheme for siting the terminal and deep water seaport. The whole project would be undertaken as a giant land reclamation exercise.*

tinctly disenchanted with the handling of Heathrow's development Every operator agreed the ugly sprawl and traffic congestion would become increasingly worse as the numbers of passengers—and the size of aircraft—increased.

It was already the world's busiest international airport. Furthermore, its complex jigsaw of terminal buildings, car parks, hangars, maintenance areas and administrative blocks were becoming noticeably more crowded every day. This situation could not continue indefinitely. The whole place was in serious danger of literally grinding to a complete halt.

Even Gatwick, London's second airport, was becoming congested at peak holiday periods. But for much of the year there was still enough room to operate—and expansion plans were well under way. Although the proposed new roads, faster rail links from Victoria station and a second runway had still to be approved (and a public inquiry held), this seemed the right direction to go.

Both the British companies fully realised that, no matter where the third airport would eventually be sited, it was unlikely to be as conveniently placed as Gatwick: there was already a good train service, it was on the main A23 trunk road between London and Brighton, and it was fully operational.

So, with these factors in their minds, officials of BEA and BOAC began a very skilful game. Not wanting to be shoved out to some far-flung airport either on the Essex coast or in the country somewhere between London and Birmingham, they decided to sneak into Gatwick by the back door.

BEA started it in March 1969 when it established a holiday charter company called Airtours and announced it would fly out of Gatwick. It was a clever move. Their fleet of refurbished Comet jet airliners—used on "package" holiday operations—at least gave the company a bridgehead at the airport which could be developed for scheduled services at a later date.

Once established, it was not long before BEA made its second major move. In early November 1970, Sir Anthony Milward, then Chairman of the airline, made it clear he wanted to buy £10 million worth of Boeing 707 jets to expand these tour operations. He also announced he was willing to talk to British aircraft manufacturers about more large planes that could be used for cut-price holidays. He received formal Government approval before the end of the month.

So the bridgehead was complete. Operating now from two major airports, BEA felt it could hardly be expected to move to a third, at least for many years. And, by this time, "Jump Jet" aviation may have arrived to make such a move unnecessary.

BOAC was not in the same happy position. It needed to spread its wings to a second airport but was unwilling to go up to fifty miles out of London to find it. So, like BEA, it attempted a back door move into Gatwick—using the recommendations of the Edwards Committee as a key.

The Committee had proposed that a "second force" independent airline should be established as a rival to the State-owned companies. BOAC's management saw the opportunity and, in March 1970, made a surprise £10 million cash offer for British United Airways.

It was a wonderful idea. BUA was responsible for 33 per cent of the total operations from Gatwick. BOAC reasoned that, if they could take over this business, any move to a third airport would be delayed for a long time. They, too, would have an expandable bridgehead for future operations.

But the scheme backfired. Mr Roy Mason, then President of the Board of Trade, discovered that secret talks between BOAC and BUA managements had been taking place for some time and that he had not been kept fully informed. So he retaliated by putting the proposed deal "on ice" because he was convinced he had not been given all the available facts.

The result was that, because of the Government's displeasure, further talks took place between BUA and the other interested party, Caledonian Airways, a Scottish company with a large slice of the trans-Atlantic charter business.

Representing British and Commonwealth Shipping, 90 per cent shareholders in the airline, was Baring Brothers, the city banking concern. They reached agreement with Caledonian and the £6.9 million deal went through. The "second force" airline came into being on 1 December 1970—and Gatwick was to be its operational base.

The Caledonian take-over included, not only a sparkling fleet of new jets, but all the ancillary equipment necessary to run a profitable international airline. It was a welcome partnership for the Scottish charter company and BUA.

But it left BOAC distinctly out in the cold, and, apparently, stuck with the inevitable—an eventual part-move to the third airport. It now seems likely that the State airline will move its expanding charter operations to the new terminal, keeping its scheduled services at Heathrow.

However, in its "house" journal—*BOAC News*—on 14 August 1970, the airline's 21,000 staff had been given the management's reluctant choice of the four sites short-listed by the Roskill Commission. Of Foulness, Thurleigh, Nuthampstead and Cublington, BOAC had selected Thurleigh.

This was the only indication that the airline had taken its pick, although senior officials were quick to explain that the journal's views were not necessarily those of the company.

BOAC News explained that commercial and operational reasons lay behind the choice. Executives had taken a careful look at all four and decided Thurleigh "best served the convenience of the greatest number of air passengers".

Furthermore, it did not require the closure of any existing airfield. It also offered access from most parts of the country without the need to travel through London and was likely to be the most profitable to operate.

These views had, in fact, been presented to the Roskill Commission by Mr Ian Jackman, a youthful 24-year-old from BOAC's Secretary and Solicitor's office. In the airline journal on 4 September 1970, he said:

"The third London airport is a piece of transport investment, one of the biggest single pieces of transport investment this country may ever see. BOAC made it quite clear that we had no local vested interest in any of the four sites. Our main concern was that the eventual choice should be operationally viable and commercially attractive. These considerations led us to a preference for Thurleigh.

"We tried not to get involved in any of the controversy. By and large, we are putting forward evidence of a commercial and operational nature that was not really in dispute."

BOAC agreed that its support for Thurleigh might be a minority opinion. But Mr Jackman told the Commission:

"We have been interested to note that, although each site has a local resistance association, only Thurleigh appears to have what we might call a local supporters' club."

This was a group called TECDA—the Thurleigh Emergency Committee for Democratic Action—which had been organised by a number of local trades union branches.

The big independent airlines based at Gatwick—Caledonian-BUA and Laker Airways—were more than happy they had their operational base south of London. Whichever of the four short-listed sites the Roskill Commission or the Government had chosen, they were likely to be unaffected.

Those operating from Luton and Stansted were not so happy. Channel Airways, which had only recently moved to Stansted from Southend, faced another move if an inland site had finally been selected.

Britannia Airways, Monarch Airlines, Court Lines (formerly Autair), all busy holiday tour companies based at Luton, were in the same position. For these two airports—only a few minutes flying time away from Nuthampstead, Thurleigh and Cublington —would almost certainly face closure to avoid dangerous flying congestion in the area.

Few airline managements will talk in detail of switching operations to a new airport. But many millions of pounds would be involved. And when would the move take place? A year after the new terminal opened? A decade, after business at the new terminal had made it impossible to continue flying from the surrounding bases? One thing is certainly clear: some of the smaller independent companies will simply go out of business if forced to uproot and go elsewhere.

At an airports conference, held in London late in 1969, papers were presented about the siting and costing of airports of the future. Discussions covered the massive expansion of passenger and air freight traffic—and the problems of airport congestion. This is fast becoming a world problem and not at all confined to Britain.

New ideas, new philosophies, revised economic calculations were hammered out and many of the experts—airport managers, planners, architects, engineers—were firmly in favour of coastal sites. They were ideal, it was thought, because they were capable of expansion either seawards or inland, if necessary.

One of the speakers was Herr Wolf Hess, son of the former Nazi leader Rudolph Hess, an engineering architect who is designing a new airport for Hamburg. He said:

"I think the possible sites that have been named are too far away from London. Hamburg's second airport is only 15 miles from the city centre and we are ensuring that, even there, the surface transport is the most modern we can devise. Delays are the bane of the modern airline traveller.

"You in Britain have a geographical problem. I accept this.

But my own personal view is that Foulness, for example, is much too far away."

Another German expert, Herr Suessenguth, chief executive of Lufthansa, called for a totally new concept in airports management. Air terminals, he said, were becoming the victim of their own size and complexity. New dimensions in aviation had brought about the problem of "super growth". More large jets were coming into service every day: air traffic had outpaced the industry's capability to handle the new generation of planes in ground management terms.

Herr Suessenguth made one suggestion which was not greeted with much enthusiasm but which is now quite likely to be accepted, at least in part, before too long: "We cannot think in terms of single companies . . . We must think in terms of doing things together, even if it means we have to give up some of our individual identities."

It is more than possible that, in the next ten years or so, airlines using the big new airports will "pool" their ground operations to make life easier for themselves and the passenger. In this way, management would be simplified too.

Furthermore, future airline passengers may find themselves provided with such services as hire cars, hotel rooms and other facilities. They might also find that parcels by air would be handled at the airport by one single company, jointly owned by all the airlines. One of the more immediate advantages of this would be a pruning of costs—and a faster service.

But, wherever Britain's future airports are going to be, one of the basic problems will be labour. People with established roots in one place are loath to move elsewhere, even if the job goes with it.

Where, for instance, would the British Airports Authority and the airlines get sufficient skilled labour to operate a new terminal at Foulness? Would staff transfer willingly from Heathrow and Gatwick? If so, they would almost certainly need help with housing costs and demand some sort of resettlement money. If not, could enough local people—or those moving into the area simply because there was a new airport there—be "educated" locally?

Airlines employ huge ground staffs and recruitment is always a major worry. BOAC, for example, feel that few of its staff would want to move to Foulness, or anywhere else for that matter.

Anyway, a move of any sort costs money. A move to a new

airport costs millions. BEA and BOAC are in a better position than most of their foreign competitors, so far as Britain is concerned. They are nationalised industries, underwritten by the taxpayers' money. But even if the Government gave them loans to move to a third airport, the money would still have to be repaid—with interest.

But with Heathrow already reaching saturation point and a similar position predicted for Gatwick in the late 1970's, future moves will almost certainly have to be made—by everyone.

As one airline chairman said: "All we seem to be doing these days is waiting for something to happen, some legislation, some new act, some new bill to spell out what the future means.

"Airlines today are almost all in the red. They are overdrawn. Their aircraft purchases, even if repayable over ten years, involve immense interest charges.

"Costs have rocketed. It now costs £10 million for a Jumbo, £10 million for a Concorde, about £7 million for the new generation of air 'buses'. With this sort of investment to be faced, with a decision to be faced on a new airport and all the facilities that would go with its creation, one cannot see airlines at least moving into a solvent position until the late 1970's at the earliest."

CHAPTER NINE

The Man in the Middle

If any one man was caught in the dangerous crossfire of the long-running Third Airport saga—someone once called it a sort of "technical Peyton Place"—he was Mr Peter Gordon Masefield, the tall, eloquent and persuasive chairman of the British Airports Authority, set up by all-Party agreement in 1964.

Stansted is one of the five airports it administers, along with Heathrow, Gatwick, Edinburgh and Prestwick. At one time he and his colleagues at the Authority's headquarters at Buckingham Gate, London, had few doubts it would become the capital's third airport—a masterpiece of planning and jet-age sophistication.

But the Labour Government's decision to set up the Roskill Commission, which eventually left Stansted off its short list, smashed these dreams. The blueprints of a new airport had to be shelved, possibly for ever.

A barrage of charges have been fired at Mr Masefield: that he regarded the Stansted development as a *fait accompli*; that he got his facts and figures wrong; that he grossly exaggerated the passenger growth at Heathrow and Gatwick in the seventies.

Many "experts" believed he had over-simplified the third airport venture as a major project in Britain's economy, and also that he over-dramatised its importance in timing. No one ever believed Mr Masefield was empire-building for the BAA. It was recognised that he was the man in the middle and could hardly step in and reveal his innermost thoughts about the situation. The experts accepted his general case that there would be air and ground "strangulation" at Heathrow and Gatwick—but not by 1974 as he had forecast.

In fact, his forecasts have been proved remarkably accurate.

Mr Masefield wrote in the Authority's first Annual Report in July 1967:

"Heathrow has already reached its night jet flight capacity, in terms of current noise limitations, and the overspill of London traffic from Heathrow to other airports in the summer of 1968 will begin to reach proportions which will be serious to the national economy.

"By the mid-1970's Heathrow and Gatwick will both be operating to capacity on all major factors and the need for a Third London Airport to accept the annual rate of increase of more than 20,000 aircraft movements and more than 1½ million passengers will be paramount.

"In May 1967 Her Majesty's Government has stated its conclusion that, of all the alternatives, Stansted best meets the need for a Third London Airport."

What Mr Masefield forecast in terms of traffic growth has—four years later—turned out to be almost precisely right. Air traffic at Heathrow, Gatwick and Stansted airports in fact increased over those years at an average of 22,000 additional aircraft movements and at 1.56 million extra passengers a year. By 1971 Heathrow and Gatwick had begun to approach capacity at peak periods and charter traffic was being turned away from Heathrow. The existing runway at Stansted in 1970 recorded an increase of 119 per cent in numbers of passengers, compared with 1969. More than 519,000 passengers were handled at Stansted in 1970 compared with only 13,000 passengers in the year 1967—an increase of forty times.

What has still not been generally appreciated is that the capacity of the existing London airports, *excluding Stansted* (on the evidence of 1971) will be exceeded in or about 1974—and that had the proposals to make Stansted into the Third London Airport gone forward, inevitably Stansted would be out of operation between 1970 and 1974 while it was being prepared for its new role.

On current evidence, without Stansted, Heathrow and Gatwick will have to accommodate between them some 500,000 aircraft movements in 1974, which is up to the full capacity of their three combined runways. Paradoxically the continued use of Stansted's existing runway will relieve the pressure by making available additional capacity sufficient for perhaps two or three years beyond 1974.

Fifty-six-year-old Mr Masefield still believes Britain might be "caught short" by not having a third airport ready in time to catch the traffic overspill, especially if Heathrow and Gatwick are further limited on noise grounds. With it would come the tremendous loss in "invisible earnings", particularly in precious dollars from North America. Would not the bustling tourists from America and Canada prefer to overfly London to the Continent and other holiday centres to avoid the disruption and chaos because of overcrowding and lack of amenities at Britain's two premier airports? he has asked many times.

No one can say Mr Masefield is a beginner in the difficult and highly technical business of aviation. He has been involved in its complexities all his working life and has held many distinguished appointments. After leaving Cambridge in 1935 he joined the design staff of the Fairey Aviation Company, but after two years went into journalism. He worked on the technical journal, *The Aeroplane*, and from 1939 until 1943 was Air Correspondent of *The Sunday Times*.

But he was fast climbing the aeronautical ladder, starting in 1943, when, until 1945, he became personal advisor (Civil Aviation) to the Lord Privy Seal and Secretary of the War Cabinet Committee on Air Transport.

Then came a spell as Britain's Civil Air Attaché in Washington and another two years with the Ministry of Civil Aviation as Director-General of Long-Term Planning and Projects. For seven years, 1949 to 1955, he was Chief Executive of BEA and, therefore, can claim to have been in at virtually the birth and adolescence of Britain's biggest airline in terms of passengers carried today.

Mr Masefield, never a man to stand still for long, later moved into the aircraft industry and became Managing Director of the old Bristol Aircraft Company, where he introduced into service the quietest large aeroplane ever built—the Britannia. With the jobs came the awards. He has been a former President of the Institute of Transport, a past President and Member of Council of the highly-respected Royal Aeronautical Society and Chairman of The Royal Aero Club.

The background speaks for itself. Although Mr Masefield has had his fair share of arguments, probably the biggest in his career came when he went to his present job and the Stansted affair blew up in his face. But he was no stranger to Stansted Airport. He had indeed first made its acquaintance in an emergency landing there in 1944 in a B-17 of the United States Eighth Air Force, shot up on a bombing mission.

He recalled: "Stansted began to be considered as an airport for London way back in 1953 when I was Chief Executive of BEA. There was a White Paper (Cmd 8902) on 'London's Airports', in July 1953, which said that "London needs three airports'. Number one it said was Heathrow, and that is right; number two it said was Gatwick, and that is right, and I would endorse that very strongly. And third, it said, that Stansted must be held in reserve as a third airport when the first two were full up.

"I said from a British European Airways' point of view that, at that date, Stansted was too far away so far as we were concerned. We certainly endorsed Gatwick as the second London Airport and would move substantial services there in due course. Stansted was probably least unsatisfactory of the other available sites. Stansted did not really come up again for ten years—although it was kept in reserve.

"These decisions, and the long-term policy over Stansted, were taken by the Tory Government which had been returned to power two years earlier. The people involved were Mr Alan Lennox Boyd (now Lord Boyd of Merton), Minister of Transport and Civil Aviation, who was succeeded in July 1954 by John Boyd Carpenter. They both endorsed the policy and so too did Mr Harold Macmillan, who was Minister of Housing.

"He set up a Gatwick Inquiry in 1954, which recommended the development of the airport in Surrey, at first with one runway and eventually with two. When that was full a third airport (probably Stansted) would be needed. So Stansted virtually disappeared from the public scene for about ten years."

Mr Masefield explained: "What happened was that Gatwick began to build up from 1958 onwards and started to take some of the load, chiefly from charter flying. Stansted was in business on a small scale, making use of its excellent long runway of 10,000 feet. But there was not the pressure on requirements at that date, although it was clearly seen that there would be pressure in the future. That is why the Inter-Departmental Committee was set up by the Ministry of Aviation in 1961. It took a great deal of evidence and in 1963 recommended Stansted out of 18 sites. The recommendation was accepted by the Government of the day and Julian Amery, the then Minister of Aviation, wrote a Foreword to the Report. In it he said that the Government accepted that Stansted was the right place for a third airport."

Mr Amery said in his foreword: "We have plans for the expansion of Heathrow and Gatwick. But the day must come when they will be working to the limits of their capacity. A third major airport will then be needed for London. We must ensure that this new airport is ready in time to take the traffic. The choice of a site for a new airport will not please everyone and the choice is limited by technical considerations. This Report discusses these in detail. It concludes that Stansted Airport should be selected and designated as London's third airport. The Government believes that this is the right choice."

Said Mr Masefield: "It is interesting, looking back, at some of Mr Amery's phraseology. He emphasised that the airport must be ready in time to take the traffic. He said, also, that the choice of a site for a new airport would not please everyone. That was some seven years ago and, I think on reflection, that he put it pretty mildly."

So, in the sequence of events, both Labour and Conservative Governments had endorsed the Stansted plan, both totally unable to forecast the stormy days yet to come. Mr Masefield has no personal grudges against the people who effectively had the Stansted plan thrown out by the Roskill Commission, or the people who lived locally, and who, by their constant campaigning, changed many people's minds.

Asked bluntly if the Stansted plan failed because a very tightly-knit determined group of people, mainly farmers with great resources, launched a massive anti-airport campaign and did a tremendous public relations job on the Government, Mr Masefield smiled and said: "Yes, I think that is probably right. I think that their protests fell on fruitful ground among some of the national newspapers who were not persuaded that Stansted was the right choice. Particularly Fleet Street, of course, tends to live around the north east of London and, therefore, there might have been some vested interests in Fleet Street which were not all that keen to see an airport in that particular part of the country. But I took quite an objective view of this when I first came to the British Airports Authority. The BAA was not in any way involved in this, right through until after the Chelmsford Inquiry of 1965. And for a year after that we had not been directly involved. I took the view that I would have liked an airport to be developed closer to London than Stansted. I saw a number of objections to Stansted in that it was in a very difficult place to get to and from. Too difficult, in fact, until the M11 motorway is completed. So I was not a very ardent supporter of Stansted as London's third airport at that date."

Mr Masefield also agreed that possibly it was not so much that the Government changed its mind about Stansted, but that its mind was changed for it. There were voices like Mr Peter Kirk, Conservative MP for Saffron Walden in Essex, and the courageous Mr Stanley Newens, a left-wing Labour MP for Epping. Mr Newens lost his seat at the last election. Both men tabled many similar questions in the House although their political views were very wide apart.

Mr Masefield's comment was: "You see this was not really a party political issue. I had wished that it would be a national issue; that it would be looked at from the point of what was right for the nation as a whole, and for our economic advantage as a nation. This, in the event, was the last thing which was discussed in the Press or Parliament. There was little debate upon what were the national needs and requirements in the interests of the national prosperity to which air transport was contributing to a large degree. But I believe that, wherever the site (and this is a very important issue in my mind), the same opposition would have grown up as that which emerged at Stansted.

"To those who are disturbed, all airports are unattractive for people who live close to them. To balance this they are attractive to shops and local businesses which do well around airports. But for dwelling houses under the flight paths we have to accept that an airport is, at present at least, a nuisance to residents. That was not peculiar to Stansted. But Stansted attracted opposition from a particular section of the population. Somewhere else would have attracted no less opposition from others.

"Where the thing went wrong, I think, was a lack of adroitness in handling it from a departmental point of view. A year elapsed after the Public Inquiry at Chelmsford. The case for Stansted was not powerfully put at that Inquiry because people, I suspect (and I must add here I was not involved in it), thought it was a pushover and they did not bother very much. The Blake Report, following the Chelmsford Inquiry, was, in consequence, unhelpful to those who wanted Stansted. And then there was a year of silence while the opponents built up their case. That, I think, was wrong. In my view what ought to have been done was that the Report should have been published, without comment, as a White Paper. Then any aspects which needed further review should have been explored in open session."

Did Mr Masefield think that Inquiries appeared to have been turned into Star Chambers and rather overshadowed their real purpose? And, another point of interest, that Mr Douglas Jay, then President of the Board of Trade, who was very much on the side of the BAA in supporting Stansted, left the Government because of this stand? Mr Masefield replied: "Stansted may have had something to do with it—but more probably not. There were, I think, other issues. The Common Market probably had a bigger bearing. But it is certainly true that Douglas Jay was very much in favour of Stansted. We took a completely objective view on the requirement from the BAA point of view. We tried to keep this stand all the way through—while understanding the local feelings which, incidentally, were not all hostile. It is often forgotten that we have an obligation to provide airport capacity; that we own and operate the airport at Stansted; and, thirdly, that successive Governments—both Tory and Labour—had declared for it. Stansted appeared to offer operational advantages—even if rather far away. So Stansted did not appear to be a wrong decision if you consider the matter from the point of view of what was best for the nation. That is what we went along with. We owned it, we operated it, it was there. And it could be developed as a step-by-step progression without causing undue noise to the local populace because—after all—there were not very many of them to be directly affected. And I think that we have as much experience of the problems of noise as most people. We know how tough these problems are."

Of criticisms that he got his sums wrong, particularly about the runway at Stansted, which some people allege was pointing in the wrong direction, and of people like Sir Roger Hawkey and Mr Lukies, who headed the preservation group, Mr Masefield commented: "Well, they set themselves up to know more about air transport than those who had been in the business for a long time. I have always avoided any personality attacks in this and I do not intend to start now. I try to be courteous and patient, even with people who attack me. Stansted and a Third London Airport was not a personal issue in any way. The issue is simply what is required for the country's trade and travel needs? The nation needs sufficient airport capacity where it can be best provided economically, and where it will best serve the national interest. Of course we have a regard for local people. But the majority are not those who live around airports. The majority are those who use air transport and gain an economic benefit from it."

At the Roskill Commission it seemed to emerge that the

British Airports Authority did not recognise the huge support for knocking Stansted on the head, particularly by the Essex and Hertfordshire Councils which included many powerful people.

Mr Masefield explained: "Yes, of course, that was so. But this was an emotional affair, emotional on local issues. It was probably, as has been suggested, whipped up by relatively few people who did not take the wider issues into account.

"Broadly, one can say that the Stansted runway was—and is —satisfactory and that a second, parallel, runway would have been equally satisfactory and that (if the long-term view had been taken) the noise level would not have been excessive. It would certainly not have been greater at Stansted than at a number of alternative sites—and much less than exists at Heathrow."

He maintained that he has always stuck to the facts. There was no emotionalism at the BAA. Although the preservationists may have been motivated by emotion he said the BAA always took the line that it wanted the best possible site available—and as near to London as possible. "We took that line all along. Show us a better site for an airport than Stansted and we shall be delighted. What we want is the best airport for the nation—taking all factors into consideration—including local interests."

Around this time the BAA drew up a points system, a League Table, of all potential airport sites. There were 48 of them. The aim was to see what would be the best in the Authority's view. Stansted was not the only possible site by any means. But it stood high and it was there with its existing runway.

Mr Masefield then spoke of the proposals to build the airport at Foulness and how Mr Bernard Clark, and later Sir John Howard, had put forward their proposals for developing an airport on reclaimed land there. He realised that it was a big scheme; a venture that could, if a deep water seaport was included, cost many more millions than an airport. In the end, the accent had switched from the airport site alone to that of an entire complex, basically aimed at providing a seaport.

Mr Masefield said: "All this was milling around behind the scenes, it did not really affect the issue. Foulness appeared to be a nonsense economically, as well as a nonsense operationally. And, in our view, Foulness did not fulfil the major requirements of an airport: that it should be operationally viable; that it should be accessible by the travelling public and that it should be acceptable to the airlines.

"All the airlines—British and foreign—have been consistently opposed to Foulness. But that was still a side issue. The main issue was what was the Government's policy? Up to the time of Douglas Jay leaving the Board of Trade the Government's policy was quite consistent for Stansted. That was altered when Tony Crosland came and felt that—in view of the political feelings (in particular the House of Lords), the whole matter should be looked at again. It was that which really swayed the issue rather than all the bellowing about in the background."

(Forty Senior Peers in the Lords had decided to back Bernard Clark's Foulness Plan. They had been so impressed by his outline proposals, his costings and his arguments in favour of using the scheme, that they formed their own Pro-Foulness Committee. They once proposed, at a private meeting, that they should develop and promote their own Foulness Bill, which was constitutionally possible, push it through the Lords and thereby present the Government with a *fait accompli*. It is interesting to note that all these Peers were not Tories, nor were they all wealthy men with industrial interests and money to speculate on any such venture.)

Throughout the third airport affair, world airlines kept a beady eye on the developments and the political issues in Britain, because it could affect their operational future. Of course, as mentioned in detail in a further chapter, both BOAC and BEA, the two state airlines, had their own ideas about how they could escape a major move to a third airport: setting up a bridgehead at existing Gatwick by creating a holiday offshoot firm or, in BOAC's case, buying their way in by a takeover of an independent airline already based there.

The airlines, it seemed, were pitifully slow in looking in depth at the growing muddle in the situation. Most of the airports being built in the world today, certainly those which handle major international operations, are no more than 15 to 30 miles at the very most from city centres, and there are examples in Paris's new airport, Amsterdam, New York and Tokyo.

Mr Masefield has an answer, perhaps not complete, for the airlines' lethargy in looking closely into the third airport situation.

"I think the airlines were a bit slow in realising the importance of the issue to them. They felt that if they sat back, the thing would 'come out all right on the night'. I think they did not fully appreciate that this was a vital issue from the point of view of the economics of air transport. Each felt that it would not have to move, that *others* would have to go. The independent airlines took the view that they wanted to stay at Gatwick, that the major airlines and all the little airlines should be pushed into a Third London Airport. If this was done, and if BEA, BOAC, Pan American and TWA all stayed at Heathrow and then, as necessary, went to Gatwick, it would not affect them this side of 1990."

"And, of course, that is pretty well true. If all the little airlines were pushed away, the big airlines could stay. But that was not the real world. That was an imaginary world which is not the world we live in. So there was a good deal of unrealistic wishful thinking in the air transport world which did not come to the surface until Roskill got going."

Was the Roskill Commission then a waste of time? "It is quite clear, of course, that the Government had to reserve to itself the decision on where the airport should be, particularly on an issue that has such political overtones.

"Mr Fred Corfield, the present Minister of Aviation Supply, when he was Minister of State at the Board of Trade, made it quite clear that the Government reserves the final decision on the Third London Airport.

"But, of course, it was difficult for any Government to go completely against a Roskill recommendation when it has been based on such detailed survey, so much time, so much energy and so much expense—not that this would have stopped the howls of anguish if, say, Cublington was selected. There was still a Stansted-type opposition. Mr Justice Roskill's job appears to have been tackled in a manner so thorough, he will have been cut aside from any charges of bias. With the time and money spent on the Commission it seemed at times that Roskill leaned over backwards to prove that he is the Complete Unbiased Man."

Mr Masefield, who has personally studied all the three million words and depth of reports prepared by the Roskill Commission, has unqualified praise for their work. Since its appointment he has never been critical of the Commission's job, except perhaps on the manner in which the short-list of four sites was evolved. This was one of the most important since the war, thrust on an independent body of experts to decide the future of a major enterprise costing many millions of pounds and involving public money.

Mr Masefield said: "I think that Sir Eustace Roskill and his team were skilled people who have taken an objective view without emotion. I was not altogether happy about their short-listed sites. We were not keen on any of them. But the Commission has done the most detailed cost/benefit study which has ever been attempted. In the final analysis, however, the question remains whether that is the right way to go about technological issues in the world in which we live? In the end, ironically, it comes down to a political decision anyway."

It is an open secret that the BAA kept its own "intelligence unit" studying all the machinations that took place behind the scenes—commercial and political. They had noted the immense Dutch interest in a possible Foulness project. They had studied the plans outlined by Mr Bernard Clark and later, Sir John Howard. They had looked at detailed plans of airports projected or being built in other parts of the world. The Authority had detailed information on the problems of surface transport and the problems of handling hourly deluges of passengers discharged from 360-seat Jumbo Jets. It had tried to forecast with accuracy, for Mr Masefield's own information, what might eventually happen. For instance, BEA had made clear that it was determined not to move to any new site wherever the Government decided upon the location of the new airport.

For all the criticism, some accurate, some misguided, which has been levelled at Mr Peter Masefield on the issue of the airport siting, he is still the man in the middle. His arms have been very much tied because of his British Airports Authority responsibility to the Government. The Authority is charged with making a profit like any commercial concern. But his view has been since the Roskill Inquiry ended that he should not get involved in the power struggles with engineers, the politicians or the rich groups who have been fighting to protect their own interests or enhance them for the future.

If one, for instance, had canvassed the four short-listed sites—Nuthampstead, Thurleigh, Foulness and Cublington—what would have emerged? Private assessments would almost certainly have come up with figures that proved more people agreed with siting the airport in their area than opposed it. But they had no effective cohesive voice. They lacked an organisation man. The BAA has always thought that the majority of opinion could not cancel out the power groups. But there was no effective way of fighting back through the BAA.

Mr Masefield has not only the thoughts of London's Third Airport constantly on his mind, he has the growing problems of

Heathrow and Gatwick. Last December (1970) an Inquiry opened in Surrey, to hear from people who are objecting to the extension to Gatwick's existing runway; so Mr Masefield has these problems too and to see that Heathrow keeps running efficiently. There is no doubt that the BAA, within the limited resources of its existing airports, has done a fine job. Meanwhile, with a further Gatwick Inquiry ahead on a second runway, his number two airport and its future expansion is still in the melting pot. Mr Masefield knows that the opponents of Gatwick's expansion are powerful. They have an "in" with the Government, certainly MPs, mostly those who live around that rich and prosperous area known as the "stockbroker belt". The situation could lead to another impasse; possibly further long delays.

Already Stansted, too, has figured in objections to minor extensions to its existing terminal, where the passenger accommodation, lavatory facilities and refreshment facilities are overcrowded because of rapid growth of charter flights.

Mr Masefield privately sees a similar situation wherever one wants to build an airport. One question he will not answer is: Where will London's fourth airport be located? (and there will eventually have to be one). Could it be Stansted? Could it be back to square one? Could the circle be completed once again?

"I am not in a position to make political judgments. That is for the Government," he said.

CHAPTER TEN

What if it Happened to You?

Imagine the crescendo of local anger, the absolute rage, if the Government—"after thorough consideration of all available sites"—decided to build London's third international airport at Orpington, Kent, only 15 miles from Big Ben. There would be an immediate stampede to lobby MPs at the House of Commons and thousands would protest march around Parliament Square.

Or suppose the chosen site was Romford in Essex, Ewell in Surrey, or Harrow in Middlesex, all about the same distance from central London? There is little doubt that the ensuing screams of complaint would reduce the Stansted and Cublington rows to the status of a mere whisper.

Even if these sites were not already development, the local residents would argue that they were much too close to the middle of London for a modern international airport. The disruption to the surrounding environment, it would be loudly argued, would be intolerable. The idea itself sounds ludicrous in Britain.

Apart from certain provincial cities—where an airport so close to the centre is possible from a planning point of view—it is unthinkable in relation to such a sprawling place as London.

Yet this is exactly what is happening in Paris where the world's first international airport specially designed for Concordes and Jumbo Jets, is being built just 15 miles from the Arc de Triomphe. Furthermore, it will be opened before the end of 1972.

Its very conception spotlights the difference in approach to new airports between the French and the British authorities—a supersonic-age variation on *A Tale of Two Cities*.

For, while London has been desperately seeking a suitable

site for its third airport for more than eight years, the far-thinking French have had theirs ready and waiting for the last 12 years.

The new French airport—at Roissy en France—has been designed as an all-weather terminal which will be fully-operational 24 hours a day. This is unlike many other international airports including London, where the screams of jets are banned for much of the night.

This has been made possible at Roissy because the French planning authorities have restricted other developments, not only around the immediate airport area—but along the approach and take-off paths. Furthermore, runways are being laid on a pattern that will keep noise nuisance to a minimum. Even the four sparsely populated villages which surround the airport—they have a total population of only 6,000—will hardly be troubled. These vital 12 years are the important margin by which Paris leads London in selecting the site for their third airport. And because of this long-term thinking the Paris Airports Authority— the French equivalent to the BAA—have had hardly a murmur of protest from the 450 owners of the 6,000 acre Roissy site. Compared with the scrappy and hastily-reached decisions that have taken place in Britain, it seems almost unbelievable that just one single houseowner was forced to move home when building began.

He was a farmer called Josef Cuypers. And, according to the PAA, he raised no objections at all. M Cuypers, it seems, was "quite happy" with the compensation he received for his small farm at Mortieres. This was not surprising for M Cuypers (and two other owners whose land he had been using for growing potatoes, cereals and beet) share a £10,000,000 pay-out for the property.

The Paris Airport Authority officials modestly admit that they have been "extremely lucky". Things worked out just the way they wanted, they explained. But there is no doubt, it was good forward planning rather than good fortune that kept the Roissy site restricted to farm land until the airport building began towards the end of 1967.

Remarkably, the site for the airport was chosen only a few weeks after the foundations were being laid for Paris's second terminal at Orly, which currently handles the bulk of the 9,000,000 passengers who fly to the French capital every year.

Le Bourget, the smaller of the two present Paris airports— and just four miles from Roissy—will gradually be phased out as the new airport develops. By comparison with London, the Paris passenger transport figures are small. More than 16,000,000 people, for instance, use London's two main airports every year. But, like the British authorities, the Paris planners realise that traffic will increase more than five-fold by the 1980's. The French estimate that 45,000,000 passengers a year will fly to Paris by 1985. And more than two-thirds of these will be handled at Roissy.

By comparison the BAA estimate that Heathrow and Gatwick will be dealing with about 80,000,000 people a year by 1977. This is why Mr Peter Masefield and his colleagues have been pressing successive governments for a quick and early decision on London's third airport. They fear that overcrowding at London's two existing terminals will force hundreds of dollar-earning flights to divert to the Continent.

Clearly, London will not have its third airport by 1977. It will almost certainly be under-way by then, but far from completion. The French have no such problems. Even before work began at Roissy the finishing touches were being made to a new motorway which sweeps under Roissy's main runway on its 125-mile link between Paris and Lille. This immediate access makes it possible to drive from the airport to the centre of Paris in twenty minutes—even during the morning and evening rush hours.

Furthermore, when the express metro link with the terminal is opened, it will take less than half that time—a quick ten-minute journey—to the Arc de Triomphe.

Paris Airport officials hunch their shoulders and spread their hands in sympathy with the difficulties faced by their British colleagues. They agree that had they not selected Roissy way back in 1957, they might have had a similar problem. They point out that if the decision had not been made then the site would almost certainly now be populated with housing and industrial developments.

The French are fortunate in another respect. The lay-out of Paris can in no way be compared with London. The French capital does not "sprawl" into surrounding towns as London does. Skyscraper flats on the Paris outskirts meet open fields as abruptly as Brighton's promenade meets the sea.

The French regard their new airport as a showpiece. They have deliberately planned it in five specific sections—with five main terminal buildings and the same number of runways. The total estimated cost after completion in 1985 has been put at around £400,000,000. This is slightly less than ten times the

BAA estimate of building Stansted. But the two are about as comparable as a pint of English beer and a bottle of vintage champagne. Whereas Stansted was a make-do-and-mend affair, with nobody really favouring the site, the plan at Roissy is to build the airport section by section as it is needed.

This will mean extra space and extra runways as and when they are required and, with the possibility of "jump jet" aviation just around the corner, they may never be required at all. The first section will be opened in 1972 at a cost of around £120,000,000. Like those to follow later, the new terminal building, an entirely circular structure with a 620 feet diameter, will be ready at the same time as the first of the 11,000 feet runways, and air traffic control building, taxi-ways and hangars.

The terminal building itself, 240 feet high, will be surrounded by a circular cluster of loading bays. Passengers will board aircraft directly through enclosed corridors linking the bays. Furthermore, at the top of the building will be car parks for about 4,000 vehicles.

This is the master plan. Each of the remaining terminal sites will be developed as needed, possibly at intervals of two or three years. The second to be finished will be used exclusively by Air France, the French national airline.

The Paris Airports Authority gleefully point out that Roissy has enough spare space to continually develop—even for aircraft which have not yet reached the drawing board stage. If longer runways are needed, for instance, it will be a simple matter of lengthening the existing ones. There will be no planning problems. Extra terminal space can also be provided without delay. All future developments at Roissy will be geared very closely to immediate requirements.

The PAA are well aware that vertical take-off aircraft could leave them with runways for which they have little or no use. And that is the crux of the matter and the reason why they are building their airport in sections. The French fully appreciate that "jump jet" airliners on the scale of the Concorde would probably require a runway of perhaps only 200-300 yards. If this happens, they argue, why build longer ones?

As the numbers travelling by air increases by several millions each year, particularly in the booming trans-Atlantic charter business, airlines will undoubtedly go where facilities are best. And France is determined to lead Europe in supplying this high-earning commodity.

Even now they are actually preserving a site for yet another airport about forty miles to the west of Paris. Airport number four, if ever needed, will be ready for the flying machines of the year 2001. The French confidently predict they can meet that schedule comfortably. Judging by what they have done at Roissy, they are probably right.

CHAPTER ELEVEN

The Modern Canutes

If the Government gives its approval to the scheme for building a massive deep-water seaport, transport complex and airport at Foulness, it will be the biggest land-reclamation exercise seen in Britain for centuries. But it will not be the first. King Charles I saw the advantages of reclaiming land from the sea and employed Dutch experts to reclaim large areas of the Fen district. In their primitive dredgers the Dutch crossed the North Sea to reclaim more than 2,000 square miles of land on our eastern coast and later constructed a Thames wall at Dagenham and effected repairs to others. King Charles was so impressed with Dutch efficiency that Cornelius Vermuyden was employed to drain the Royal Park at Windsor and later reclaimed 70,000 acres of land at Hatfield Chase.

The Dutch, from absolute necessity, had been winning back land from the sea for over eleven centuries. They, more than anyone else, had been fighting a running battle with their constant enemy. Their problem was a simple alternative: push back the sea or leave the whole of Holland to be swamped. Periodically over the centuries, the North Sea has devastated large areas of the Netherlands, the most recent case being in 1953 when nearly a third of the country was under water. Over this long period it is not therefore surprising that they have become such great constructors of water defences and land reclamation. Self preservation is a mighty incentive. After so much effort in building dykes and reclaiming so much land the great flood was a devastating blow. But the Dutch decided that it was also the last straw and were determined that it would never happen again. So they started an enormous long-term programme to change the entire pattern of sea defences throughout Holland. Within a matter of weeks and before most of the water had subsided, a scheme was completed and the work started almost straight away on what is now referred to as the Coastal Protection Scheme. In general terms, this is a system of great solid dykes, dams and barriers running the whole length of the western border of Holland to protect it from the North Sea. The system is, to build a massive wall and series of locks to control the water and the various outlets of the Rhine. These waterways are some four and five miles wide in places and the effect of building this massive wall from north to south, under which there is also a motorway, is to shorten the coast line of Holland by some 250 miles. In other words, cutting out the coastal defence which was, prior to 1953, necessary on either side of the waterways dividing up the delta.

Everyone has heard of the Zuider Zee, but few people in Britain appreciate that this area is still being reclaimed. The effect of this reclamation project has been to add some 400 square miles to the area of Holland—an indication of what can be done in this field of engineering technique. The policy of the Dutch is not to use up valuable land but to create it.

The expertise of the Dutch in this field can probably be best explained by how they yet again reclaimed their land from the sea after the coastal dykes were bombed in four places by the Allies in 1943. The massive sea walls were breached at four points and the water poured in. More than 40,000 acres were flooded with salt water and the population fled for their lives. The breached walls widened and assumed formidable proportions. As the tide flowed in and out, the flooding became worse and worse. While the war was still going on, there was no question of repairing the damage. The only question was would it ever be possible to yet again push back the North Sea and close the gaps.

But as soon as the Allied troops landed in Holland and before the last Germans had been pushed across back into their homeland, engineers in the liberated zone set about mending the damage. At first, they were unable to reach the breaches as the area was full of land mines and there was not a single boat to be found. It now seems fantastic, but it was only after somebody discovered a rowboat on the mainland that it became even possible to inspect the blasted walls. The sea had not been idle and the damage was much worse than had been feared. Every day the tidal channels had penetrated further and further inland and the breaches had become wider and wider.

But a start had to be made and the first dredger was raised

from the bottom of the sea inlet where it had been submerged since the bombing. The Allies gave every assistance within their power and Switzerland sent in wooden sheds, Belgium granted credit for the purchase of stone, dredging equipment was sent from Britain and Belgium. Although the task was not seriously undertaken until the north of Holland was liberated—on 5 May 1945—work started with a vengeance once the mines had been cleared and the powerful dredging fleet, belonging to a consortium of contractors which had been carrying out the Zuider Zee project, sailed out to the south-west.

Within a few months, it seems there were no fewer than 320 units of floating equipment at work: 14 suction dredges, 140 barges, 17 large tug boats, about 80 landing craft and freighters, 19 large and small floating cranes, 50 to 60 drag-lines and bulldozers, and no less than 70 barracks were built to accommodate the labourers. The breaches were now of almost two year standing and, because of erosion, the damage was over 3,000 yards.

After several unsuccessful attempts to close these enormous gaps in the country's sea defences, the Dutch decided to scuttle large but otherwise useless ships and the enormous Mulberry pontoons which had been given to the Netherlands by the British Government. This was only a temporary repair but it proved effective enough to enable suction pumps to begin pumping the water back into the sea.

In just one year the Dutch were able to win back 40,000 acres of valuable agricultural land—a massive achievement by any standards. Over the last fifty years the Dutch have reclaimed no less than 550,000 acres of the Zuider Zee, a clear indication of what can be done given the problem.

There are, of course, many other examples of how man's ingenuity has won valuable battles against seemingly hopeless odds.

One of the more interesting events took place in the United States in 1830 when a group of settlers wanted a loan to develop their village in northern Illinois. The bankers refused on the grounds that "none but fools would locate a human settlement in that swamp". The village of fools went ahead anyway and the small settlement is now called Chicago.

Even today there is still plenty of reclamation work going on all over the world. One American dredging company built an entire island in the Pacific as a bomber base for the United

States Air Force. The British Army remember Dunkirk as a small port. Few knew little about it until the massive evacuation took place in 1940. Very soon it will be one of France's biggest transport complexes with docking facilities for the world's biggest ships. The French have pushed back the sea in their determination to put Dunkirk into the same money-spinning position as Rotterdam.

In the south of France, Marseilles will soon be more than five times as big a port as it was ten years ago. A massive reclamation project is well underway at the Gulf de Fos.

This, too, will be able to take 500,000 ton tankers, bulk carriers and container ships that shipping experts think will be sailing the seas in only a few years time. These projects are enormous, quite as big as anything needed by Britain at Foulness. Both have full government backing and financial support. By the mid-70's the financial return is expected to begin offsetting the vast capital outlay involved. A senior official of the Marseilles Port Authority summed up the attitude of all other European expansion ports: "This is a race. Nothing short of that. If we finish first—we get the business.

"If we are second, we get less business. If we do nothing—we get nothing. Our development scheme, like the others in Western Europe is not just a construction exercise. It is an economic necessity.

"Unless we improve and expand we will be out of date. Being out of date in this business is like being old, alone and living on a pension. You have few friends."

But the pushing men of Marseilles intend to have plenty of friends. Like Dunkirk, its northern counterpart, Marseilles is being subsidised by the French Government to the tune of 60 per cent of the estimated £100 million capital expenditure. The reclamation area in the Gulf de Fos is about the same size as Paris.

In 1978, completed reclamation will enable the port to handle around 170 million tons of shipping. In 1966 it handled 63 million tons. New industries have already staked claims to valuable port-side sites. Four huge oil companies are building refineries. Petro-chemical companies are moving in. So is a huge fertilizer concern. In all, 30,000 new jobs will be available.

In Western France, at Le Havre, an ambitious reclamation scheme under the leadership of M Maurice Thieullent, is well under way. The local business leaders are attempting to woo foreign industrialists to the area in an effort to create new wealth. They are succeeding. Renault has already arrived along with other large firms. They are building on 25,000 acres of "new land" that was under the sea only a few years ago. The port area itself is increasing in size month by month. In 1966 it handled 31 million tons of shipping. By the middle of the 1970's that figure will "at least" have doubled.

The aim in every case is to compete with Holland's fabulous Europoort project. All the others are large, impressive and far-sighted. But the object lesson can be learned from Rotterdam's showpiece port, already the world's largest and soon to be bigger still.

All the world's major oil companies, some of the biggest car manufacturers, petro-chemical industries—and all the ancillary and service companies—have pinned their faith in Rotterdam. All are investing millions of valuable pounds in the progress and success of Europoort. They have little need to worry. They know their money is secure because it is going to be difficult—if not impossible—for anyone to catch up in the foreseeable future.

The growth of Rotterdam can best be judged by annual income to the Dutch Government from the port in levies and taxes alone. In 1950 it reached £70 million. By 1965 it had shot up to a remarkable £285 million. By 1970 it was well over the £300 million mark—and still growing!

So big, in fact, is Rotterdam in relation to anything we have in Britain that large ships—loaded with goods for this country—have to discharge them at Europoort. Then, smaller British vessels chug across the North Sea to bring home the parcels in smaller packages. The increased expense, in terms of higher transport charges, is self-evident.

But one of the most amazing aspects of all these massive European reclamation projects is the speed at which they are carried out. It seems that no sooner has a dredge completed a "fill" than the surveyors move in, closely followed by the excavators and the construction workers.

As we have said, reclamation was a necessity to the Dutch. But they have now developed it to the point of being almost a fine art. They know all the pitfalls and the drawbacks. They also know the advantages—and Holland's increasing prosperity is living proof of that.

One splendid example is the magnificent Schiphol Airport at Amsterdam, opened in April 1967. Few of the millions of airline passengers who fly in and out every year realise that it was once part of a giant 45,000 acre lake.

Haarlem Lake, scene of a bloody sea battle between the Dutch and Spanish fleets in 1573, was the ideal spot for the new airport. So, the dredgers moved in, the water disappeared, and an airport was built—13 feet below sea level. The land is almost identical in character to Foulness.

Further up the Dutch coast is the new town of Lelystad. It stands at the end of a 16-mile dyke. A few years ago it was part of the North Sea. New industries are arriving, tarmac roads are being constructed, houses are being built. Just outside, farmers are now ploughing rich new land and cows graze in the fields where fish once swam.

It is all an impressive sight. Few in Britain have ever seen it. Few, in fact, have ever realised just how much reclamation work has been done in Holland—or how much is taking place in other parts of Western Europe.

But the result is the same. Britain is being left further and further behind in the economic race, so far as port facilities are concerned. The Dutch have cornered the lion's share of the market. The French and the Belgians are attempting—and will almost certainly succeed—in capturing a smaller share for themselves. Meanwhile, British ports continue to decline.

Meanwhile, the only successful Canutes in history stride on. They are pushing back the sea and intend to keep pushing. Land is much more valuable than water.

CHAPTER TWELVE

Bernard Clark—Mr Foulness

Bernard L. Clark is a quiet, modest, grey-haired consulting engineer with a small, but thriving business in Westminster and an unpretentious family house in suburban Surrey.

Until the summer of 1967 few people outside his clients and professional colleagues had ever heard of him; most of the public still do not know his name, although they have probably seen him on television a few times and, perhaps, heard him on radio.

But Bernard Clark is fast becoming recognised as the Cecil B. de Mille of the construction business. The famous film maker whose productions were invariably on epic scale—"with a cast of thousands"—could not think in "B" movie terms.

Neither can Mr Clark, but his production ideas are even larger and certainly more enduring. His latest scheme, for instance, is to create an entire new county for England by reclaiming 600 square miles of the Thames Estuary.

"It would mean thousands of new jobs and new homes," he says enthusiastically. "It would also mean we could control the flow of the Thames and stop, once and for all, the flood dangers to London."

He is also the man who first hammered out the scheme for building London's third international airport on reclaimed land at Foulness, off the Essex coast. It was far better, he decided, than ravaging the lush, productive acres around Stansted.

When his plan was first published, in the *Daily Telegraph* on a hot day in June, 1967, many people, including fellow engineers, laughed at the scheme and thought he must be suffering from a mild bout of midsummer madness.

"There are two things wrong with it," said one of London's most eminent engineers. "The first thing is the cost. The second is that the Government has already decided to build the airport at Stansted. You might also add that it is too far from London."

But, within a few days, the possibilities of the scheme dawned on several leading City business interests, development and construction companies. Members of Parliament were made aware of it—they were all sent a copy through the post.

By the end of the year financiers had moved in, there was a Foulness lobby in the House of Commons and the House of Lords; the Noise Abatement Society, the Stansted Preservationists and others had all taken up the "Let's Build at Foulness" cry.

It developed from there. He added a massive deep water dock to the airport plan, spoke in terms of a British Europort, lectured, talked, broadcast and gave Press interviews about the idea. Finally, even the most doubting of Thomases saw the clear commercial possibilities. Bernard Clark's grand design was finally off the launching pad.

But it was not a master plan born of overnight flashing inspiration: no sudden moment of genius: no wave of a fibre tipped pen across a map of Eastern England.

True, the idea of putting an airport on the sand flats was new. But the concept of developing the Thames Estuary into a gigantic industrial-residential complex had been mulling over in his mind since the 1930's. It had been nurtured by years of experience and planning; strengthened by the sight of developments eating their way into the heart of rural England.

The airport rows, the Public Inquiry, the hundreds of Commons' questions, the very idea of laying miles of concrete at Stansted, provided the spark that lit the Foulness fire.

But it all began fifty years ago when, as a boy, Bernard Clark found himself fascinated by the wonders of construction; the pyramids, dams, bridges, harbours, large buildings. "They were a veritable magnet to me," he recalls nostalgically.

"This interest has remained with me throughout my life It is my good fortune really that I enjoy my work so much it is also my main hobby."

So, when other boys read their comics, their adventure stories, followed their own idols, Bernard Clark was finding out about the men whose drive, imagination and expertise made Britain the world's leading maritime and industrial nation.

"What became very interesting to me in the early days was the fact that these small islands of ours have produced most of the great men of the past—the men who built, designed and invented all the things which have made our present-day life possible.

"These heroes of mine range through from Metcalfe to George and Robert Stevenson. Smeaton, the great harbour and lighthouse builder; Rennie; Bolton; Watt, a great man to whom we owe our electricity and even computers today.

"Telford, the bridge builder. Christopher Wren. I wonder what he really was. Was he an architect, which we all call him? Or was he an engineer? Certainly his St Paul's Cathedral today would be a great engineering feat. Yet this was constructed some 300 years ago, only rivalled in recent years by the tall, steel-framed, reinforced concrete buildings.

"Brunel, the great railway man. Brassey, probably one of the biggest contractors who ever lived. In his time he employed some 50,000 men—as many as most big industries today. He was a man, who, by his engineering skill, contributed so much to the relief of the British Army at Sebastopol.

"He was the man who built railways across the world—across Russia, China, India, throughout Africa and in South America.

"What about Wheatman Pearson, later Lord Cowdrey, a great man who developed the oil industry and did so much work in Mexico and South America generally?

"All these men came from Britain. It is important to appreciate that these people, and a handful of our own bankers, like the Rothschilds, controlled nearly all the railway building in the world, outside the United States, between 1840 and 1870—and a large share of transport development half a century later. The scale of work as individuals is gargantuan and undeniable.

"Their efforts, and those of other Britishers like them, changed the history of the world. Their efforts made European nations the great industrial powers they are today. They are at least, and often more, important than the political figures of history."

With these men as his idols, it is not difficult to see why Bernard Clark thinks on such a grand scale. He was fed on massive constructions; he grew on the drive and enterprise of Britain's engineering masters, the stories of their armies of navvies travelling from one gigantic enterprise to another.

Even today he still looks with wonder at their performances and points to the economic effect they have had on our daily lives. "A great many major constructions were undertaken in the nineteenth century. The great North Sea Canal which was cut through Holland created the city of Amsterdam.

"This spurred the people of Manchester into building their Ship Canal. This was an enormous achievement: miles of water-way built by hand at a cost of nearly £20 million. It was done ninety years ago but it has made Manchester the second city in this country. It can take ocean-going ships from the Atlantic into the heart of Manchester and we all know what development has taken place because of that.

"It is interesting to consider this: the construction of the Manchester Ship Canal involved the digging out of almost the same quantity of material that would be required for the Foulness airport-docks project. And yet this was done nearly ninety years ago—and by private enterprise."

What annoys and saddens Bernard Clark now is that the British are no longer thinking in these terms. Now, he says, "the red tape syndrome" operates. Everything needs years of Parliamentary debate, argument and, even then, probably indecision and the shelving of a worthwhile project.

"And this sort of negative thinking is taking place at the same time as our formerly great assets—our docks, industrial centres, transport systems—are getting old and out of date. We need a new burst of life to put us back at the top. New thinking, big thinking. Otherwise we are going to be left behind. So far behind, in fact, that it will become impossible to catch up.

"That is why a scheme like Foulness is so important. It would mean an enormous transport complex, the best in the world. It would give us the best and most modern docks, the best airport. And round this could go the refineries, the factories, the large industrial centres that would, for once, be facing in the right direction—Europewards.

"That is where our future lies and, I think, deep down, everyone knows it. It is no good selling cars and heavy machinery to Europe if we have to transport them from the Midlands and the North. Put the works on the dockside, so to speak. This cuts transport costs down to a minimum and ensures quick delivery to the waiting ships and speedy delivery the other end.

"If we do not do this we are going to be in serious trouble. Look at Europe already. Take Holland and their Europoort. Magnificent. Take other places and you will find the same thing happening. Everyone else seems to have forged ahead while we have been standing still. The amazing thing is, we are still wondering why. More people ought to go to Europe, not just for a holiday, but to look at what is happening there industrially. They will get the shock of their lives."

In his efforts to gather as much information as possible about reclamation techniques, Bernard Clark has made scores of visits to Holland, consulted Dutch specialists in the various techniques, enlisted the aid of Dr Jan van Veen, considered the "father" of modern reclamation methods.

During his trips he has become friendly with economic advisors to the Dutch Government, men who know what can be achieved when the planning and the know-how is there. Some are regular visitors to his Victoria Street, Westminster office, and many a lunch at the nearby St Ermin's Hotel has finished with lists of complicated mathematics being scribbled on the menu.

He has also enlisted the aid of the Netherlands technical colleges which have special faculties devoted to reclamation problems, delving into the questions of tides, land disturbance, ballast, stonework and silting.

The result has been that Bernard Clark is—by British standards—an expert on the complications of reclaiming land. He knows the problems, the difficulties and the costs. The advantages are obvious to everyone.

But, although he admires the Dutch immensely and is long in their praise, nobody can ever accuse Bernard Clark of being either apathetic or lukewarm in his attitude towards Britain. When he says "I am completely patriotic", he is not being even slightly cynical. He means it. He is the complete patriot.

As his studies of Foulness have progressed over the last few years, he has spent a small personal fortune in his search for information. His own staff, who could have been well occupied on other matters, have been drafted to help on the scheme; he has hired aircraft to fly over the area many times so he could film the sand flats at high and low tides, in smooth and stormy weather —in all sorts of conditions.

Furthermore, his many trips to Holland have not been cheap. The travelling costs of giving more than 200 lectures, spreading the gospel of Foulness, have been substantial. The costs of issuing more than 20,000 documents of one sort or another would alone send lesser enthusiasts scurrying for cover.

But, although his firm has benefited indirectly from all the publicity, Bernard Clark has not received a penny from anyone for his labours. And he says he does not want anything.

"This is a national asset we are dealing with—land lying in shallow water just off our shores. It is for everyone and everyone should share in it. In fact, one thing which upset me more than anything else was the way some firms jumped on the Foulness bandwagon because they saw a quick profit in it.

"The whole venture is too big for quick profit, too much for one firm—or group of companies—to handle. It needs careful planning and organisation. A meticulously phased programme of development is essential. Then, it will be found, there will be enough work for everyone. The sheer size of the scheme makes it the biggest thing that has ever happened to this country.

"It is a marvellous opportunity. We should seize it with both hands while we can. If we miss the chance now, we will never forgive ourselves in the future. Nor will our children."

CHAPTER THIRTEEN

Foulness—the Advantages

Passengers in today's superbly built aircraft are almost certainly in less danger than people who drive their cars through central London during the twice-daily rush hours. But one fact is inescapable—planes do sometimes crash.

Occasionally, these crashes mean the loss of many lives, not only of passengers and crew, but of people on the ground as the plunging aeroplane cuts its trail of disaster through homes or factories.

Furthermore, the vast majority of these accidents happen as the plane is either landing, taking off, making its airfield approach or initial climb. A recent world-wide survey (excluding Russia and China) showed that of 136 crashes, no less than 101 occurred during these critical periods.

This fairly conclusive evidence seems to indicate that constructing airports in or around built-up areas is sheer lunacy if the situation can possibly be avoided. The likelihood is that, at some time or other, a plane will crash either in the district surrounding the terminal, or the area over which it makes its approach.

In this event, people will almost certainly die—just because they happen to live or work on the approach or take-off paths. Imagine what would happen, for instance, if a Jumbo Jet crashed on central London while making its final approach to Heathrow. Think of the possible death toll if a Jumbo loaded with nearly 400 passengers, dived into a block of skyscraper flats, or busy Oxford Street. Quite literally, it could mean the lives of thousands. And what a public outcry there would be!

Righteous politicians, particularly the publicity seekers, would demand courts of inquiry, Royal Commissions, a total ban on flights over the capital, the abolition of all Jumbos. And they would be supported by a large section of the outraged public.

Many countries have already realised the possibility, and have made sure their international airports are as far removed from residential and industrial property as conveniently possible. No less than 18 of the world's leading airports are sited in exclusive positions like this.

But Heathrow is not so well placed. Landing and take-off paths mean the aircraft have to travel either directly over congested London, or, coming the other way, over Windsor and Maidenhead. Gatwick is better placed—slightly—from this point of view, although Redhill, Crawley and Reigate are only a stone's throw away.

Given the inevitability of a major crash at some time, and the statistics indicating at what point in the flight it is likely to happen, it seems reasonable to give the pilots a "crash area" as free of obstacles as possible. Only one of the four sites short-listed by the Roskill Commission provides this major advantage—Foulness.

An airport built on reclaimed land, like that proposed for Foulness, has natural crash-landing facilities. The shallow water surrounding the 14-square mile man-made island would limit the possibility of fire on impact. Further, the water itself would be a form of soft "cushion" to prevent the plane breaking in two, or as often happens when crashes occur on land, disintegrating altogether.

By landing in water, the aircraft is, therefore, more likely to remain in one piece and be fire-free.

In these circumstances, passengers will almost certainly survive because the impact will be far less than a land crash. Finally, modern planes are designed to float in water for quite a long time, so the airport's rescue boats would have time to pick up passengers and crew before the unfortunate plane sank. And, lying in shallow water, salvage operations would be easier, thereby giving experts a better chance of discovering what went wrong.

Several airports around the world are already built on reclaimed land, including La Guardia in New York, Rio de Janeiro and Hong Kong. All these terminals, with landing and take-off areas over the sea, have had planes crash into water. In most cases, the majority of passengers have escaped. One particular instance, which happened in Hong Kong, is worth recalling. A Caravelle crashed into the sea alongside the runway. There were

100 people on board. One died and several were slightly injured. The remainder escaped.

What, one wonders, would have happened if the same crash had happened at Heathrow, Gatwick, Stansted, Cublington, Nuthampstead or Thurleigh? Or any inland site, for that matter. It is a reasonable bet that the death and injury toll would have been much higher.

And what of the noise problem? Any airport administrator knows that a terminal that can be used 24 hours a day in all weather is a far more profitable proposition than one with restrictions on night flying. Heathrow already collects more than £7 million a year in rents and landing fees from foreign airlines. This figure would be far higher if these restrictions had not been imposed. The same applies to Gatwick.

But for the sake of people living in the area, who like everyone else need their sleep, noise limits had to be imposed. If the third airport had gone to Stansted, the same would have applied there —and at Cublington, Nuthampstead and Thurleigh. This, in turn, would have restricted the profitability of the new terminal.

But not necessarily at Foulness. With landing, take-off and approach paths over the sea, people, even in local Southend, would hardly be troubled. What is more, some of the "stacking" areas could be over water, thereby cutting out most of the noise nuisance to people sleeping on land.

Noise, therefore, is a vitally important factor in deciding the location of any new airport. As Sir William Hildred, formerly Director-General of the International Air Transport Association, pointed out in a foreword to the feasibility study commissioned by the Noise Abatement Society:

"Noise, of course, is something the other chap does—the beatnik with transistor radio, the bachelor trombonist in an upstairs flat, the moron simulating virility by the exhaust of his motorbike, the neighbour's garden mower on Sunday, the whine of a jet, the laughter of nit-wits, street drills—in modern life is full of hideous cacophony—but do not let us make it worse."

About the advantages of Foulness, he explained: "This is not primarily a financial problem. As with every vast undertaking, finance must enter in. But there are considerations far deeper than finance—noise, accessibility, preservation of rich soil, international acceptability, but mainly noise."

Mr John Connell, the Society's chairman, commented: "Noise destroys all that is good in civilisation. It adversely affects health of mind and body, retards recovery of the sick and reduces the capacity to learn and the quality of the work done. It is vitally important, therefore, that London's third airport is built on a site where the least possible noise nuisance is caused by day or by night."

The feasibility study had proved "beyond a shadow of doubt" that Foulness was ideal from the noise "and every other point of view". It also stated: "For Foulness, the noise contour includes no populated areas and lies over reclaimed land or over the sea. From the noise point of view, Foulness is incomparably better than Stansted . . . The ten-mile circle of approach on Foulness is mainly over the sea and it will be possible to organise air traffic control so as to "stack" over the sea".

The study, comparing Foulness to Stansted, produced five more important factors in favour of the coastal site. Visibility records taken over a six-year period showed Foulness had a "far better record". It had no early morning mist, and during the winter months, there were periods when poor visibility was only half as frequent as at Stansted.

Prevailing winds at Foulness were in line with the runways and would have no adverse effect on operations. Weather conditions—temperature, rain, sunshine and snow—were better than Stansted's. On average, it would be open to traffic eight more days a year. "A further factor standing very much in favour of Foulness is in the fact that the lower ambient temperatures of the sea at take-off will considerably increase the potential payload of aircraft."

Apart from the amenity organisations, which have almost all supported Foulness, the coastal sites biggest and most influential backing has come from the local authorities. Whereas every other suggested site has been ferociously opposed by local government, the reclamation scheme has been welcomed with open arms.

The Essex County Council have backed it from the start and Southend was not far behind. The Government-financed South-East Regional Planning Council has given its official blessing and so has the Greater London Council.

These, by any standards, constitute a powerful political lobby. But add the weight of the Port of London Authority, that enormously influential body of opinion known as "The City", with its massive financial resources, and the Foulness supporters' club becomes mighty indeed.

The reasons are simple: the coastal area, and the districts for some miles inland, are ripe for development. They need to be developed, or the whole region is in danger of suffering from slow economic strangulation.

A large international airport on the coast would, inevitably, need surrounding service industries: hotels, motels, garages, warehousing facilities, cargo and supply depots. The 100,000 or more people who would either drift, or be drafted, into the area would need housing. That means shops, stores, schools, hospitals.

In short, the whole inland area would begin to grow and assume some identity with the airport as its umbilical cord.

Southend airport would, of course, be forced to close. But the site is owned by the local council who would develop it for industry or housing, or both. The town's hotels, at present reliant on the summer tourist business, would have a guaranteed supply of customers all the year round: local shops would flourish: the whole resort would become a sort of district capital.

The Essex County Council, which joined the fight against Stansted, would also benefit. An airport at Foulness, with its network of surrounding development, would mean more money in the county rates purse. Furthermore, the authority could stop worrying about attempting to attract industry to the district.

Within the context of its much larger parish, the regional planning authority sees the terminal as a blessing from heaven. It has long been conscious of the population drift away from east London to the more popular west side. There are lots of reasons for this, mainly economic. East-side industries were old and dying; the west-side companies were new and expanding.

The result was plain enough. The east areas were becoming drab and depressed while the west progressed and looked healthy. Anything to stem this situation, indeed reverse it, would be welcome. The magnet of an international air terminal provided a ready-made answer.

On their part, the amenity associations—the Council for the Preservation of Rural England, for example—see Foulness as the only way of stopping civil aviation requirements turning England's green and pleasant lands into a jungle of fumes, neon signs and concrete.

The Port of London Authority's support is not so much based on its liking for a coastal airport as the proposed deep-water dock complex that should be built alongside it. The dock, like the air terminal, would be on reclaimed land. Both jobs could be done together, probably cheaper than separately. They would, after all, have to live and work together, so to speak. It therefore seems sensible that the two should start and develop hand-in-hand; one was more likely to then appreciate the problems of the other.

Before Mr John Davies, Secretary of State for Trade and Industry, made his Commons announcement on 26 April 1971, that Foulness was to become the third London airport, there had been enough "leaks" to the Parliamentary Press corps to make the official statement almost an anti-climax.

Outlining his reasons for making the choice, Mr Davies recalled that even the Roskill Commission—which, by a majority vote favoured Cublington—agreed that Foulness was the best site on grounds of planning and environment. This was a key factor in the decision: these considerations "were of paramount importance".

He made no secret of the fact that the coastal site would be many millions more expensive than an airport at Cublington, or any other inland area, but the price was worth paying to save the vast tracts of premium farmland, lovely countryside and the settled communities.

Total cost of providing an airport at Foulness would be around £500 million at 1971 prices. But this included the vital road and rail links with central London. The first planes would take-off from the new terminal in 1980, he forecast.

His suggested timing for the new airport was an important part of the whole political lobbying which has surrounded the Foulness argument since 1966. Sir John Howard's group has always claimed the finance was readily available, the planning work already completed, and that construction could start almost immediately.

But, while Mr Davies made it absolutely clear the Government was willing to consider the possibility of private money being invested in the scheme, the 1980 date indicated that the Cabinet has decided not to rush its fences. In other words, no private group of contractors were going to get their hands on such a prime site and develop it for speedy commercial gain—even if it meant delaying the building work.

Sir John had previously indicated that the first runway would be completed by 1972–3 if he got the contract. Although the bait must have been tempting for the Government, the cautious Mr Davies decided to wait several years longer to make sure the results were right. Indeed, the Government's detailed planning will not be completed until 1973. The bulldozers, dredging equipment and ancillary gear will not arrive until later. There will be no haste in this scheme.

The Cabinet considered the reversal of the Roskill recommendation very carefully before making any public announcement. But its final conclusions were that the economic penalties to be paid for choosing such an expensive and distant site would be compensated by speedy motorway access and efficient airport operations.

Indeed, the reluctant airlines would, in effect, be forced to move to Foulness by stricter Government limits on aircraft movements from Heathrow and Gatwick. This would have the added effect of reducing noise nuisance to people living near the two present terminals.

"It will be open to the British Airports Authority to arrange (landing) charges between its airports to stimulate traffic at Foulness. On these assumptions, which differ from those made by the Commission, the new airport can be expected in time to make a proper return on capital invested there, though it may not become self-supporting as quickly as one at an inland site," Mr Davies told the Commons.

He also made it quite clear that he understood the interested development groups were mainly concerned with the potentially much more lucrative seaport complex than the airport. He temporarily pacified them by explaining that all concepts would be closely considered—including the seaport and surrounding industrial complex.

Meanwhile, the Government was going ahead on the basis that the airport would be built with a seaport tagged on the end, rather like another piece of a jigsaw puzzle.

Although Mr Davies did not spell it out in precise language, the Cabinet must have had another factor at the back of its mind when making the decision about Foulness.

Only 300 people live on the 7,000-acre island. And the majority of those are pensioners, many in the eighties. There is a noticeable shortage of youngsters. Only 15 children, for instance, attend the local single-classroom school. In 1960, there was twice that number.

So Foulness, as a population centre, is dying anyway. Even the pensioners, robust, healthy individuals though they are, cannot live indefinitely. In the present circumstances, there will be nobody to replace them. A flat island landscape, easily transformed into a viable airport terminal, with a populace whose average age is over 60, was a natural political choice. Very few votes are lost by troubling and displacing the least number of people.

The few islanders who will be left when the construction equipment rolls on to the thirty square miles of Maplin Sands, will have to be rehoused. It will be a terrible blow to most of them, particularly since they have lived most, if not all their lives, in their little clapperboard houses, listening to the boom of the exploding shells from the Ministry of Defence artillery testing ranges nearby; tending their precious little vegetable gardens and watching the flocks of Brent geese and wildfowl.

The whole island will eventually change. The handful of tenant farmers—who, like everyone else at Foulness—till their soil by courtesy of the Defence ministry which owns the whole area, will also move. They will have difficulty finding similar farms at the low rents charged by their military landlords.

As an isolated community, Foulness has a character of its own. Inevitably, everyone knows everyone else. Many of the islanders are related. Some have never left the place in their lives.

Take the Shelley family, for instance. Mr Harry Shelley is over 80. He lives quietly with his two elder brothers, Charlie, aged 83, and Ernest, aged 82. The three old men are cared for by their younger sister, Miss Bertha Shelley, aged 70. None of them have been away from Foulness for a total of more than two years.

Down the road lives Mr William Clark, aged 78, and his wife, Mrs. Amy Clark, aged 76. Both were born at Foulness and cannot remember the last time they left the place. Neither can many others.

But the youngsters leave, as soon as they are old enough. Apart from the almost non-existent social life, there is another, more pressing reason—work. If you do not work for a local farmer or at the artillery ranges you have to travel outside or be unemployed. There are no other jobs. The average wage in the area is around £13.50 a week, hardly incentive for an ambitious youngster to stake his future on the place.

Obviously, as soon as work starts on the new airport, the artillery ranges will be forced to close and move elsewhere. The military is reluctant to make the change and this is hardly surprising. They have been in the area, punching their heavy shells into the sand flats, since 1805. A home which has been "in the family" for such a long time is difficult to leave—especially since the conditions at Foulness cannot be reproduced elsewhere.

Still, move they must, probably at a cost of around £25 million. This was the figure given the Roskill Commission by military experts. How they arrived at that sum, nobody knows, particularly since they did not know where they were likely to move, when, or how. The units will now almost certainly split into fragmented groups and dispersed to different parts of the country.

The only other casualties caused by the arrival of the airport will be the geese, ducks, swans and other wildlife. They will have to find new breeding grounds, possibly further up the east coast towards the Wash. Their disturbance has angered naturalists who pleaded that the quantity and variety of birds in the Foulness area was a great natural asset which should be left intact. They also pointed out that it would be dangerous to fly aircraft in the district because of the dangers of birds flying into engines or windscreens.

But the Government decided it was worth the risk. It also decided that Foulness was a better long-term proposition than Cublington, or any inland site. One felt that even the tiny population of the island were in reluctant agreement.

The enormous, and highly professional, propaganda machines, established to promote the sale of Foulness as the only realistic site for the third airport, had cost around £1 million by the time the government made its official announcement in April 1971.

Most of this money came, not from private individuals incensed at the possible rape of their beloved countryside by the concrete of runways and the petroleum fumes of screaming jets, but from big business interests: entrepreneurs who saw the coastal sand flats as prime commercial property, a virgin piece of real estate capable of being developed into a highly lucrative transport complex.

Furthermore, these developers were quite prepared to spend another £1 million, if needed, because, whilst the speculative outlay would be high, the potential returns would be even higher, particularly if the government gave permission to build a giant deep-water seaport alongside the new air terminal.

In this respect, the whole propaganda effort was somewhat misleading. Because, whilst the entire campaign was slanted at shaping public opinion in favour of a Foulness airport, the real, almost unspoken object, was aimed at winning approval for the docks scheme.

The City reasoned that if the government decided to locate the airport on the reclaimed Maplin Sands, it would be only a matter of time before the seaport was tagged on the end, so to speak. Both would be reclamation projects and together they would provide the south east with the world's greatest integrated transport complex: it would be a natural marriage.

Whether the money was spent in vain, we have yet to discover, but now, having given Foulness airport its official blessing, the government will come under increasing pressure from the development companies to link both projects. The City says that, whilst London needs a new airport, Britain also needs a new seaport to take the world's biggest ships and thereby challenge Rotterdam, at the moment the world's biggest and busiest port.

But the airport scheme did not start in such a grand or well-organised manner. It was on 21 June 1967 that the *Daily Telegraph* published a report that Mr Bernard L. Clark had a plan to build the controversial airport on the Maplin Sands by reclaiming land from the sea.

More than a handful of people, including many alleged experts, were convinced he was quite mad, and his proposals unworkable. The first people to see some sense in it were officials of the Noise Abatement Society, particularly Mr John Connell, its full-time chairman.

The Society had, some years earlier, suggested Foulness as an airport site, but their idea was to build on the island itself, not as a reclamation exercise. Mr Clark was told they would support him and had, in fact, already had an initial meeting with the firm of Covell, Matthews and Partners, architects and planning consultants.

The Noise Abatement Society's primary aim was to reverse the government decision to build the airport at Stansted. Not surprisingly, Mr Clark was also immediately supported by the North West Essex and East Herts. Preservation Association— the official anti-Stansted protesters. They saw it as a viable alternative and a possible means of saving their farms and picturesque villages.

In a way it was a curious situation. On the one hand the Labour government, having already voted in favour of Stansted and seemingly determined to proceed with their choice were almost immediately badgered by MPs on both sides of the House to, at least, have a look at the Foulness plan.

On the other, Mr Peter Masefield, chairman of the BAA, and a Stansted supporter, was continually reiterating, publicly and privately, that the government should give permission for work to start straight away.

He incessantly stressed that Heathrow and Gatwick were already turning away dollar-earning aircraft at peak summer periods, and that both airports would reach saturation by 1974. The only site which could be operational by then was Stansted.

This was true, but the public was not in favour and, in some respects, neither were the government planners, who only regarded the choice as "the best of a bad bunch".

The government was, therefore, in something of a cleft stick. It was forced to take heed of Mr Masefield's warnings, yet, at the same time, it could not totally ignore the cries of protest from its own supporters in the Commons. Finally, there loomed in the background the real chance that the House of Lords would throw out any Bill promoting Stansted. The Peers had already indicated their implacable opposition to the scheme.

It was not really surprising, therefore, that the opponents of Stansted greeted Mr Clark's plan with great enthusiasm. They saw it as, at the best, a real alternative, and, at the worst, a possible way of delaying the destruction of the Essex countryside.

As Mr Clark said afterwards: "I was staggered at the almost immediate response. Phones in my office never stopped ringing. Groups invited me to speak to them and outline the whole scheme. Businessmen, bankers, and all sorts of interested parties wanted to arrange meetings to talk things over."

Besides the MPs and Peers—who had all received a copy of Mr Clark's proposals through the post—one of the first to show real interest was Sir John Howard, head of John Howard Ltd., civil engineers, and a former chairman of the National Union of Conservative and Unionist Associations.

His firm had been involved in many large-scale constructions. It built the foundations of the Severn and Forth Bridges, had wide interests in many commonwealth countries, and once won a £40 million contract from President Nkrumah of Ghana to build a deep-water harbour.

The two engineers had several meetings and Sir John decided to form a company to promote Foulness airport and produce its own scheme. He asked Mr Clark to join the venture, but his offer was declined. Mr Clark reasoned that, as a professional consulting engineer, he could not be party to such a commercial venture, even though Sir John suggested that a member of the Clark family could serve on the board as Mr Clark's nominee.

By this time the Foulness band-wagon was gathering speed, and the government's case for Stansted was being shot full of holes as more and more evidence on its unsuitability emerged Furthermore, determined Peers had formed their own pro-Foulness committee in the House of Lords, and Mr Clark was having a series of regular meetings with them to map out a campaign to foster and organise support.

In fact, Foulness received so much support that, before the end of the year, there was a veritable glut of plans, all broadly similar and differing only in matters of detail.

In addition to Mr Clark's originating scheme, there emerged one from Sir John Howard. He launched his ideas at a lavish Press conference at London's Waldorf Hotel—ideas which turned out to be amazingly similar to those of Mr Clark.

Later, another Press conference was held. At this, the feasibility study for Foulness, commissioned by the Noise Abatement Society, was published. This impressive 78-page document, carefully prepared and splendidly illustrated, also proposed the same site as Mr Clark's first scheme.

But it was brilliantly compiled, and included a comparative study on Foulness in relation to Stansted—which was still the government's number one choice. Its closely-reasoned argument were very clear: Foulness was the right place for an airport, and Stansted was manifestly wrong.

Together, these studies constituted a welter of evidence in favour of the coastal site. Fleet Street and television had already taken up the cudgels on behalf of the Essex villagers, and the government's case grew daily more tattered. It was now quite clear that the Cabinet, already seriously split over Stansted, could no longer ignore the claims of the reclamation planners.

What is more, the House of Lords lobby was making itself felt. The Peers had sounded-out the City to discover whether financial support would be forthcoming if Stansted was abandoned in favour of Foulness. They told the government that financiers were prepared to back the scheme with hard cash.

Then came the proposal that, more than anything else, was responsible for forcing the Wilson administration to think again.

Only a few weeks after the devaluation of the pound, Mr Clark published his grand design—a massive seaport and airport complex on the Maplin Sands.

Influential financiers and developers immediately saw the commercial prospects of the venture, and even more pressure was put on the government to reconsider. Now, however, not only was it coming from the Press and public and parliament, but from the City itself—and the troubled, and powerful, Port of London Authority. The PLA, its ancient docks in a sorry state, welcomed the possibility of having a prestigious new port to rival Rotterdam's showpiece Europoort. Such an acquisition would mean the Thames could be used by shipping on a round-the-clock basis, instead of the three or four hours a day at present allowed by the tides. The long-term financial gain was self-evident.

The strength of the argument was irresistible, and, understandably, the government relented. Mr Douglas Jay, a firm supporter of Stansted, left the government, and his position as President of the Board of Trade was taken by Mr Anthony Crosland. Mr Jay's antipathy towards Britain's entry into the Common Market was an obvious contributory factor.

On 20 May, Mr Crosland announced, to cheers from both sides of the House, that Stansted was to be at least temporarily reprieved whilst a Commission investigated the timing and siting of the third airport. Shortly afterwards Mr Justice Roskill was appointed to head the inquiry.

Meanwhile, private enterprise had not been idle. Sir John Howard had busied himself by forming the Thames Estuary Development Company (Tedco) the previous February. He had enlisted the co-operation of the PLA, Shell UK, Rio Tinto Zinc, John Mowlem and Company, and the Southend Corporation, who all had representatives on the board.

In addition to providing the Roskill Commission with a mass of evidence in favour of Foulness, Tedco has also built a £300,000 hydraulic model of the scheme to assess tidal and flooding factors should the airport and seaport finally be built on the coast. In all, it has spent more than £500,000 proving the project's feasibility.

The support of the Port Authority is an important part of Tedco's plan, mainly because the PLA, by special Act of Parliament, has extended the limits of its jurisdiction to include the Maplin Sands. Therefore, any port built at Foulness would come under its aegis. Furthermore it would, presumably, have a say in who was to build the port.

Southend Corporation, as a member of Tedco, has invested about £100,000 to promote the Foulness plan, mainly on the basis that, as a town so close to the scene, it could influence policy more as a member of the group than as an outsider.

Mr Archibald Glen, who, before he retired as Town Clerk of Southend, was the town's representative on the Tedco board, was quoted in *The Times* as saying that he regarded the investment, not as speculation, but as a "research exercise".

Tedco obviously considers it stands a fair chance of landing at least part of the Foulness contract, both for the airport and, if it is ever built, the seaport. But it is not the only group interested in the development.

In June 1969, the Thames Aeroport Group (Tag) was formed with 15 major companies represented on its board. These include Rio Tinto Zinc Investments, Myton and Grand Metropolitan Hotels. Having spent more than £100,000 on promoting its scheme, it wanted to join Tedco in a 50–50 partnership. Tedco refused, on the basis that shareholdings should be proportionate to investment.

The Tag board is fearful of the possible consequences of the PLA involvement with its rival group, but the Port Authority argues that Tag has not asked for any assistance from them.

Both groups are fully aware that, despite the large sums they have already invested, there can be no guarantee that, at the end of the day, they will receive anything in return. The government might, for instance, put the whole operation out to tender. In which case both contenders could be pipped at the post, possibly by American companies which have been keeping a close eye on events, or the massive Dutch dredging companies which have for long expressed more than a passing interest.

In fact, it is difficult to see who else could carry out the actual reclamation. It is a Dutch speciality, and few, if any, other nations have the dredger capacity or the know-how. The Dutch companies have English-based subsidiaries which have been involved in the Foulness promotion exercise.

When, after more than two years of listening to and assessing evidence from all sides, the Roskill Commission announced its majority recommendation in favour of Cublington (Wing), all but the totally politically naive realised that an airport at Foulness was very much on the cards.

Cublington was cheapest to build and easiest to operate, the Commission explained. But, in the final analysis, the total cost of producing and fitting-out a major international airport is about £4,500 million.

Therefore, the few million pounds saved by building Cublington was very small beer in relation to the damage which would be inflicted on the beautiful Vale of Aylesbury. So Cublington was never really in with a chance of becoming the third airport.

Even so, the locals took no risks. They organised the Wing Airport Resistance Association (Wara) and employed the slick American public relations firm of Burson Marstellar at a cost of £3,500. Within days, the firm's professional opinion-moulders were busy in the pubs and clubs of Fleet Street arguing the Wara line with air correspondents and environmental reporters.

The resistance group, claiming 62,000 members, was financially assisted, not only from local subscriptions, but from funds collected to fight a similar battle should Roskill have selected Thurleigh or Nuthamstead, the two other inland sites which had been shorlisted.

One of the subscribers was Sir John Howard, who lives only five miles from Thurleigh. He has always insisted that his contributions were made privately, not as the head of Tedco. This is reasonable because his home would have suffered had Thurleigh been selected.

Every Sunday morning for weeks, even in the thick fog and treacherous ice of the 1970–71 winter, the ploughs, tractors, muck carts and combine harvesters of Wara members trundled their mechanical protest around the winding and narrow lanes surrounding the proposed airport site.

They even had their own song, written, sung and played by jazz singer Cleo Laine and her husband, bandleader Johnny Dankworth, who live in a splendid detached house only a few miles from Wing.

But the protests were not just voicing objections to Roskill and his unpopular choice. They had a positive side: build the airport at Foulness. The most commonly seen poster stuck in almost every window read: "Roskill no, Buchanan yes," a reference to Professor Colin Buchanan's lone pro-Foulness voice on the Commission.

What is more, and not unexpectedly, the local politicians made their voice heard. And with great effect. Mr Stephen Hastings, the hard-working, hard-talking conservative MP for mid-Bedfordshire (his constituency links both the Cublington and Thurleigh sites) had formed a committee of back bench MPs within days of Roskill announcing his short list. By the time Cublington was announced, he had collected the signatures of more than 200 MPs.

The committee was an important behind-the-scenes influence and, because of its close association with a similar group in the Lords, headed by Lord Molson, chairman of the Council for the Preservation of Rural England, formed a powerful pressure group within parliament.

These two groups were, again, opposition with a positive aim; not only were they determined to get Cublington rejected by the Government, but to make sure it decided on Foulness. Lord Molson's role, besides whipping up support in the Lords, was also to rally the leading conservation organisations behind him.

In January he, and the chairman of five preservation groups wrote a letter to *The Times* objecting to Cublington and expressing their support for Foulness.

So, by the end of February 1971, Mr Heath's cabinet was faced with a veritable barrage of powerful propaganda, all seeking to convince him that Roskill's £1,200,000 bill had been a complete waste of money and that the coastal site was the right choice after all.

There was Tedco and Tag; the Lords and the Commons; Wara; Burson Marstellar and the Press; numerous preservation societies; influencial interests in the City, particularly among the development and banking fraternities. All fighting on the same side, they constituted a formidable army, and one which even the most immovable of governments dared not ignore.

On the other side, opposing Foulness, were a handful of people living on the Army's top-secret seven-acre island, some military brasshats, and naturalists who feared for the future of the Brent Black Geese and other wildfowl which are prolific along the Foulness coast.

Finally, inevitably, Mr Heath and his cabinet voted in favour of the Essex coast, and Mr Davies, Minister of Technology, announced the decision to a largely delighted House of Commons.

The following day disconsolate Ministry of Defence officials showed parties of journalists around the island and, at the same time, Tedco announced it was negotiating with British Rail on the possibilities of establishing a jointly-financed train service between the airport and King's Cross. The object was to run a 35-minute service between the two points and thus overcome the one big snag in siting an airport at Foulness—access.

Within hours of the Commons announcement rumours sprang up in the Southend, Shoeburyness and Foulness areas about land speculation deals, about how the Post Office had installed a 3,000-line telephone cable link to the island six months earlier, and about how private investors "in the know" had made killings buying up parcels of land and local properties. The validity —or otherwise—of the rumours was, of course, impossible to establish immediately. But they were persistent enough to provide plenty of food for thought.

Some of Peter Masefield's more Controversial Comments

1 MASEFIELD (I.T.V. 20.6.67 *This Week*—Stansted For And Against).

>On the occasions—and there are about 20 per cent of occasions—when the wind is from the east, all the aircraft landing at Sheppey or Foulness would be coming right over London.

Comment: *If this rule applies, then when the wind is from the north-east, aircraft landing at Stansted will come right over London.*

2 MASEFIELD (I.T.V. 20.6.67 *This Week*—Stansted For And Against).

>45 NNI is not an intolerable figure—it's perfectly reasonable.

Comment: *It was accepted at the Inquiry that this was an intolerable figure. Further, Mr Masefield says that a much larger NNI is unacceptable to the Medway towns vis-à-vis Sheppey.*

3 BAA Handout "An Airport for the 70's—What the New Stansted will mean to you".

>From the noise point of view more people would be affected if an airport was built at Foulness or Sheppey rather than at Stansted.

Comment: *Government White Paper Cmnd. 3259 Paragraph 15 page 30:*

>*Relative extent of noise nuisance at various sites. (All jet traffic at an SMR of 64 movements an hour.)*

Stansted	*1*
Sheppey	$\frac{1}{4}$
Silverstone	$\frac{1}{2}$
Heathrow about	*20*

>*And Paragraph 48 page 15: The noise problem would be relatively small (Thames Estuary sites).*

4 MASEFIELD (*Evening Standard* 12.6.67—Interview with Max Hastings).

>Sheppey would interfere with far more people's lives (than Stansted).

Comment: *As for "3".*

5 MASEFIELD (Letter to the *Evening Standard* 16.6.67).

Sheppey could be used if the nation was prepared to face disturbance to the larger communities who live around there.

Comment: *As for "3"*.

6 MASEFIELD (Letter to *The Times* 28.4.67).

(Sheppey) . . . more displacement of houses and people . . . than at Stansted.

Comment: *As for "3"*.

7 MASEFIELD (*Evening Standard* 12.6.67).

Jet noise is at its worst right now. In the next few years we expect it to get better rather than worse.

MASEFIELD (In The Public Eye—BBC Radio 8.3.68).

Of course existing noisy engines will be with us for a long time. (Mr Masefield said later at a Dunmow meeting that jet engines would not get quieter for twenty years.)

Interviewer

Are we going to have the noise problem getting worse before it gets better?

MASEFIELD

To a small degree.

Timing

8 *Kirk* (ITV 20.6.67—*This Week*).

At what date do you propose to start work on the second runway (at Stansted)?

MASEFIELD

About 1974.

MASEFIELD (Letter to the *Daily Telegraph* 19.2.68).

Sir Roger does the country a disservice when he suggests that there is plenty of time to spare for further long inquiries and that there will be no need for a Third London Airport by 1974 . . . Stansted alone can be made available by 1974 to meet traffic needs.

BAA Handout "An Airport for the 70's—What the New Stansted will mean to you".

By perhaps 1974 the traffic will have increased to necessitate the provision of the second runway.

MASEFIELD (Statement 7.2.68—reported in the *Daily Express* 8.2.68).

Stansted must be ready to absorb some 3 million passengers in 1974 and large increases in succeeding years.
MASEFIELD (At the Institute of Transport 4.3.68).

London's Third Airport would be needed by 1974—would not now be ready before 1976.
MASEFIELD (Letter to the *Daily Telegraph* 19.2.68).

Stansted alone can be made available by 1974 to meet the traffic needs.

9 MASEFIELD (*Evening Standard* 12.6.67).

It'll just be a matter of planning to add a little to each existing village, which is already happening anyway.

Comment: *At a conservative estimate 100,000 people are expected to come into the area.*

10 MASEFIELD (*Evening Standard* 12.6.67).

Stansted—using 5,000 acres of very thinly populated countryside.
MASEFIELD (Chart presented at The City of Westminster Chamber of Commerce 17.10.67 and later published in *The Director*).

Stansted—Approximate additional area required for four runway airport 4,000 acres.

Comment: *15,000 acres is a realistic figure given by the NFU for four runways and industrial development, etc., involved, and this is without any new town which almost certainly would be required.*

11 MASEFIELD

In the Chart mentioned above there is no mention given to planning.

Comment: *(The Government White Paper—Paragraph 67 page 19). In the Government's view the strongest of the objections to Stansted is on regional planning grounds.*

12 MASEFIELD (Chart presented at the City of Westminster Chamber of Commerce 17.10.67 and later published in *The Director*).

If the airport was at Foulness or Sheppey the effect on ATC at Heathrow and Gatwick would be considerable.

Comment: *(Inspector's Report—Stansted Inquiry Paragraph 19 page 55):*
On air traffic control at Sheppey the Technical Assessor said: "There is no conflict with Heathrow or Gatwick, a point which was accepted by the Inter-Departmental Committee".

13 MASEFIELD (Press Conference 12.6.67).

> Stansted—the distance to London was reasonable, though access could be better.
>
> MASEFIELD (In *The Public Eye*).
>
> Stansted is much too far away (from London) at 34 (miles).

14 MASEFIELD (Addressing the Engineering Section of the British Association at Bristol 6.9.55).

> Nor could one neglect the need to improve transport between the airport and the city centre. The overhead, aerodynamic, monorail, running on pneumatic tyres at more than 180 mph had a great potential.

Comment: *This is what should be done from Foulness then distance won't matter.*

15 MASEFIELD (Town and Country Planning Association 1.12.55—he was then President of the Institute of Transport and Managing Director of Bristol Aircraft).

> Suggested that a monorail service held the answer to the problem of providing quick transport in areas where road and rail services were congested.

Comment: *As for "14".*

16 MASEFIELD (Letter to the *Financial Times* 19.2.68).

> A fundamental requirement of air transport operations is that air services shall be operated with the maximum of safety and the minimum of delay.
>
> MASEFIELD (Report on speech at Bournemouth 9.10.62).
>
> Agreed that safety is not the prime consideration in the air—"Safety does not come first, otherwise we should never get out of bed". (*The Times* comment 10.10.62—the point they were making was that . . . the aims of safety must be kept in balance with those of the widest possible use of the air.)

17 MASEFIELD (Letter to *The Times* 30.12.60).

> As one who uses the air continually, both as a pilot and as a passenger, I believe that alarm and despondency is out of place on our ATC procedures. The real issue is not that the air is congested but that—as yet—not enough aeroplanes use London's airports to make them pay. Hence the enormous landing fees quite out of proportion to those in force elsewhere.

Extensive development, not only of air transport but also of properly controlled executive and business aviation, during the next few years, for which there is plenty of air space available, will, we may hope, go some way towards remedying this economic problem.

Comment: *Compare his comments on landing fees now he is in charge of BAA.*

18 MASEFIELD (Letter to *The Observer* 24.12.67).
The nation cannot afford a loss of around £20m. a year on Foulness or Sheppey, even were the airlines prepared to move to such places which is exceedingly doubtful.

Comment: *Airlines are prepared to move to Foulness (?).*

19 MASEFIELD (Report in *The Times* 13.11.54 on article in BEA Magazine—then Chief Executive BEA).
The decision to develop Gatwick is "good news for our future expansion". Gatwick meets all requirements for an "overflow" airport.

Comment: *Time has proved him wrong?*

20 MASEFIELD (Interview Max Hastings *Evening Standard* 12.6.67).
"I don't believe there are more than a handful of people involved at the heart of this thing."

Comment: *(Inspector's Report—Stansted Inquiry Paragraph 115 page 26).*
North-West Essex And East Herts. Preservation Society formed for the purpose of opposing the airport, it represented at least 13,300 people, probably more.

21 MASEFIELD (The *Daily Telegraph* 29.12.67—Report on Statement 28.12.67).
"Not a chance of Government revising its plan."
No Comment!

22 *The Times* 29.3.60.
Because of the impending transfer of Bristol Aircraft Ltd., to the newly formed British Aircraft Corporation, a number of changes have been made in board membership and executive appointments of the Bristol company.
Mr Peter Masefield . . . (has) resigned from the board.
With his six years' experience as chief executive of BEA (49–55) Masefield may well be another under consideration—Chairman BOAC.

Chairman Air Transport Licensing Board—name mentioned as a possible first.

Parting with Bristol was "friendly all round".

23 24.8.55. Addressing Town And Country Planning Summer School.

Helicopter—probably take at least ten years before helicopter could become a satisfactory commercial vehicle.

Vertical take-off with fixed wing aircraft . . . for some time the noise factor would probably rule it out of city centres.

Some political promises to Stansted preservationists

1 Extract from a letter from the Deputy Prime Minister and Foreign Secretary (then Mr R. A. Butler, MP for Saffron Walden) to Sir Roger Hawkey, Bart., 11.9.64:

> "*I am glad to make the position clear as I see it. So far an expert inter-departmental committee has recommended Stansted Airport as the Third London Airport on technical grounds. No account has yet been taken of the human considerations, the living conditions, the amenities and the agricultural character of the area involved . . . What has got to be done, now, and done by your committee, is to put the case as I have described above to the Public Inquiry which is promised for next year. I pledge myself that this Inquiry shall be genuine and that the findings should be respected . . . I have the Prime Minister's Assurance that no final decision has been made, and will not be made until after the Public Inquiry next year.*"

2 Extract from a letter from the Prime Minister (then Sir Alec Douglas-Home) to Mr R. A. Butler, 13.9.64:

> "*You have been telling me of the problems which arise over the choice of a Third London Airport following upon the report of the inter-departmental committee set up to consider this question. The committee's recommendations in favour of Stansted were derived from a study of the operational and other technical requirements of an airport. The Government recognises that whilst these technical considerations carry weight, many other considerations arise such as the disturbance of local life in villages and towns in the neighbourhood, loss of amenities and loss of good agricultural land. The Government, therefore, decided that the public inquiry should take place some time next year and I am aware that you would like this not to be too long delayed. It is important that this inquiry should be thorough and genuine, and I want to make quite clear to you that the Government has in no way made its final decision, and cannot do so until it receives the report of the inquiry.*
>
> "*It would therefore be fully understandable if your constituents concentrated upon preparing their case for the inquiry meanwhile. I further say that not only has the Government not made*

its final decision but that the door is not closed to suggestions of alternative solutions if they are put forward. This is in fact a long-term matter which can be decided only after a thorough public investigation has taken place."

3 Extract from a letter from the Ministry of Aviation to Sir Roger Hawkey, Bart., 16.11.64:

"The Public Inquiry is unlikely to take place before the latter part of next year and the detailed terms of reference have not yet been settled. The main purpose of the Inquiry will, however, clearly be to consider the objections which may be raised to the expansion of Stansted into a major airport for London. The Government intend that the Inquiry should be a full and fair one and that no unnecessary restrictions should be placed on the nature of scope of the objections. In particular it will be open to objectors to give evidence of the greater suitability of any of the alternative sites mentioned in the Inter-Departmental Report (Cap. 199) or any other sites of which they have given previous notice."

4 Extract from report of resumed conference on proposed Third London Airport attended by representatives of the Ministries, County Councils, Urban Councils, Rural Councils, Parish Councils and preservation societies at Harlow on 10.12.64.

The conference opened with a statement from Mr D. A. Lovelock (Ministry of Aviation) which ran as follows:

"The new Government is committed to a full and fair inquiry on the same terms as those declared by the last Government. No sort of decision will be taken on Stansted until after the inquiry. If the inquiry comes down against Stansted, alternative sites would again be considered, and one of these selected."

5 Extract from a letter from the Minister of Housing and Local Government (then Mr Richard Crossman) to Lord Molson, 10.8.65:

"Roy Jenkins and I are fully sympathetic to the need to have a public local inquiry at which there can be a thorough discussion of the impact of an international airport on the neighbourhood before the Government reach their decision. The inquiry will in fact be held under an Inspector appointed by me, and the terms of reference (which I hope will be published soon) will make it clear that the Inspector has to report not only on local objections to the proposal but also on suggestions made for alternative sites.

"Roy Jenkins and I do feel, on the other hand, that this inquiry is not a forum in which it would be profitable to discuss the question whether there is need for a third airport for London.

This is a question which raises highly complex technical points of a character which cannot effectively be discussed at a local inquiry and issues of policy which really must remain matters for decision by the Government.

"With this reservation I believe that the inquiry will enable all the important issues to be fully ventilated. While the subject of the inquiry is the international airport itself, and not the question whether a new town should be established in the vicinity or other measures taken to accommodate the additional population generated by the airport, these will be matters which objections will no doubt mention as part of the possible effect on the neighbourhood. You may be sure that the Government will not take their final decision on the airport without regard to the likely consequential effects on this part of Essex."

6 Extract from a letter from the Minister of Aviation (then Mr Roy Jenkins) to Brigadier T. J. P. Collins (Chairman of the Planning Committee, Essex County Council), 24.9.65:

"I have considered very carefully, and at length, the representations made to me at our meeting on Monday, 13 September about the proposals to develop Stansted as the Third London Airport.

"I have concluded, however, that the objection to extending the terms of reference for the public inquiry in the way you suggested is too strong and that the original terms of reference must stand . . . However, there were some aspects of the deputation's case on which I was able at the meeting to give reassurances, which will, I believe, help to allay local misgivings to some extent. I will repeat these assurances. First, it was represented to me that, unless the terms of reference covered the question of timing, it would be possible for Ministers to make the inquiry a formality by refusing to follow up indications in the Inspector's Report that an alternative site was preferable to Stansted, on the pretext that the necessary study and survey of the alternative site would take too long and development of the airport could not wait. I assure you that this will not occur. If the outcome of the inquiry is that another site is to be preferred to Stansted, this will be followed up and it will not be ruled out for lack of time to study and survey the site.

"This brings me to what is, I believe, the most important aspect of this matter. The deputation told me that a refusal to meet their request would tend to give the impression that the Government's mind was made up, that the inquiry would be only a formality and that the development of Stansted would be pushed through regardless of the outcome. I can give you the firmest possible assurance on this. My mind is not closed to

proposals for alternative sites. It is for this reason that the terms of reference of the inquiry, unlike those of the Public Inquiry of 1954 into the proposed development of Gatwick Airport, allow objectors to propose alternatives. My only concern is that the site which is eventually developed as the Third London Airport, whether Stansted or elsewhere, should be the best and most suitable that can be found. In reaching a decision full regard will be paid to the effect which the development will have on the selected locality, not only in the immediate future but in the long-term; and it is in examining this aspect that the Public Inquiry will be particularly valuable."

7 Extract from a letter from the Ministry of Housing and Local Government to the North-West Essex and East Herts. Preservation Association, 2.12.65:

"The purpose of the inquiry is to obtain the information and views without which the suitability of Stansted as against other possible sites cannot be properly assessed; and the Inspector's function will be to hear and report on the objections to the development of Stansted, together with the representations about possible alternative sites. I can assure you that the exclusion of recommendations is not intended to prevent him from expressing a clear view on the objections to Stansted, or on the desirability of further consideration being given to alternatives, and he will be completely free to do so. It will, however, be appreciated that, since the inquiry is one into objections to Stansted, it would not be open to him to make a positive recommendation for the use of another site instead, since that would then tend to prejudice any further inquiry which might prove necessary relating to that other site.

"The Association can also rest assured that the fullest weight will be given to the Inspector's report and his conclusions, and he is being asked to express these in clear-cut terms."

8 Extract from the election address of Mr Michael Cornish, Labour candidate for Saffron Walden at the General Election:

"Some in the constituency are worried about Stansted, the proposed third airport for London, and I should like to quote the Ministry of Aviation statement of December 10th: 'The new Government is committed to a full and fair inquiry on the same terms as those declared by the last Government. No decision will be taken on Stansted until after the inquiry. If the inquiry comes down against Stansted, alternative sites would be considered and one of these selected.'

"I stress that the outcome of the inquiry is not a foregone conclusion."

9 Extract from a letter from the Eastern Regional Board for Industry to the Ministry of Housing and Local Government, 14.10.65:

> "*It is the firm opinion of the Regional Board that it would be wrong to develop Stansted as London's No. 3 airport and as a new town, since such a scheme would take land of high agricultural value, in fact some of the best in the country. It is understood that in selecting a location for the airport a basic requirement is that it should not be more than one hour's travelling distance from London. Travel developments in the next few years, e.g. monorails and helicopters, are likely to bring within an hour of London localities such as the Lakenheath/Brandon/ Thetford area—an area which, unlike Stansted, seems most suitable in many ways for the sort of development and population increase of 100,000/150,000 at present proposed for the Stansted district. Lakenheath air base possesses the longest runway in the country and is fog-free on average on all but three days per year. The provision of the airport and new town in this locality need only affect land of much poorer quality and would seemingly fit in with industrial needs too.*"